CHALLENGING PRIME NUMBER PROBLEMS

Gerald Patterson

PREFACE

The twin primes and the Goldbach conjectures are two important prime number problems in number theory. The author has been researching these two famous problems for many years and now presents the results of his labour in this book. It is hoped that the ideas in this book would inspire the reader to think further about the nature of the prime numbers and possibly lead to new ideas and solutions.

Happy reading!

Gerald Patterson, PhD
July 2019

CONTENTS

1 INTRODUCTION

Prime numbers are evidently the building blocks of the integers or whole numbers, which are either prime numbers, or, products of prime numbers whereby they are known as composite numbers such as 15 (3 x 5), 21 (3 x 7) and 25 (5 x 5), etc. A prime number, except for 2 which is the only even prime number, is an odd number which is divisible only by itself and 1, e.g., 3, 5, 7, 11, 13, 17, 19, etc. There are now several outstanding unsolved prime number problems in mathematics, viz., the twin primes conjecture, the Goldbach conjecture, Polignac's conjecture and the Riemann hypothesis, the most important being the Riemann hypothesis apparently.

The twin primes conjecture posits that there is an infinitude of primes separated by 2, e.g., 11 & 13, 17 &19, 29 & 31, etc.

The Goldbach conjecture states that every even number is the sum of two primes, e.g., 8 = 3 + 5, 10 = 5 + 5, 12 = 5 + 7, etc.

Polignac's conjecture postulates that there are an infinite number of cases of two consecutive prime numbers (including the twin primes) with difference n [n = 2, 4, 6, 8, ...].

The Riemann hypothesis asserts that all the non-trivial zeros of the Riemann zeta function (a non-trivial zero is the point in the series of the Riemann zeta function wherein all the positive terms and the negative terms in the series cancel out each other resulting in zero) **only occur** when the real part of the Riemann zeta function is 1/2 - these non-trivial zeros being indicative of the pattern of the distribution of the prime numbers.

As the author has spent much more time researching and working on the twin primes and the Goldbach conjectures and is more familiar with them, this book would cover only these two prime number problems.

2 A FEW SOLUTIONS FOR THE TWIN PRIMES CONJECTURE

PART 1

Theorem: - There is an infinitude of twin primes pairs.

Argument 1:-

Lemma: By Georg Cantor's definition of infinity something that is undeniably smaller than an infinite quantity could also be infinite. Cantor, who had worked alongside Hilbert, defined infinity as the size of the never-ending list of counting numbers (1, 2, 3, 4, 5, 6, 7, 8, …). Anything that is comparable in size is equally infinite.

By this definition, the number of odd counting numbers, that would intuitively appear to be smaller, is also infinite. It could be shown that the quantity of counting numbers and the quantity of odd numbers are comparable as we could pair off each counting number with a corresponding odd number, as follows:-

$$1 \quad 2 \quad 3 \quad 4 \quad 5 \quad 6 \quad 7 \quad 8 \quad 9 \ldots$$

$$\downarrow \quad \downarrow \quad \downarrow \quad \downarrow \quad \downarrow \quad \downarrow \quad \downarrow \quad \downarrow \quad \downarrow \ldots$$

$$1 \quad 3 \quad 5 \quad 7 \quad 9 \quad 11 \quad 13 \quad 15 \quad 17 \ldots$$

If every member of the counting numbers list could be matched with a member of the odd numbers list then the two lists must be the same size. As the list of counting numbers is infinite the list of odd numbers must also be infinite.

The primes (with the exception of the number "2" which is prime by definition only) and twin primes are also odd numbers and belong to the list of the odd numbers. As the list of the odd numbers is infinite the list of the primes and twin primes must also be infinite. (In fact, every odd number (in the infinite list of odd numbers) is either a prime number or a composite of prime numbers (i.e., it consists of prime factors), the primes being the "atoms" or building-blocks of the odd numbers (the primes could also be regarded as the building-blocks of all the whole numbers or integers, including the even numbers - vide the Goldbach conjecture, which, it appears, virtually all mathematicians believe to be true, all the even numbers with the exception of 2 are each the sum of two prime numbers - all the even numbers with the exception of 2 are also each the product of prime numbers (composite).) The twin primes could be found scattered all over the infinite list of odd numbers. All this means that just by the expediency of enumerating the infinite odd numbers, we could "generate" an infinite number of primes and twin primes. With regards to the twin primes, it is as if the innumerable 'twin primes' fruits are waiting to be plucked from the infinitely big 'odd numbers' tree.

Argument 2:-

Lemma: The development of the self-similarity concept has brought fame to Mitchell Feigenbaum, who has worked in the Los Alamos Laboratory in the early 1970s. The "distinguished presence" of the twin primes amongst the prime numbers, as described in the chapter, is analogous to the "self-similar mathematical pattern or structure" (which is the shape of the fig-tree itself) of the various parts of the fig-tree, i.e., its trunk to bough section, bough to branch section, branch to twig section and twig to twiglet section, in Feigenbaum's famous fig-tree example, and, such self-similar mathematical pattern or structure, or, fractal characteristic, could also be found in other aspects of nature, e.g., waves, turbulence or chaos, the structures of viruses and bacteria, polymers and ceramic materials, the universe and many others, even the movements of prices in financial markets, the growths of populations, the sound of music, the flow of blood through our circulatory system, the behaviour of people en masse, etc., which have all spawned a relatively new and important branch of mathematics with wide practical applications known as fractal geometry, which has been pioneered by Benoit Mandelbrot. As a matter of fact, self-similarity or fractal characteristic could be regarded as the fundamental mathematical aspect found in practically everything in nature including the numbers such as the prime numbers and twin primes which are the subjects of our investigation here, and, this new branch of mathematics, fractal geometry, besides having a great practical impact on us also gives us a deeper vision of the universe in which we live and our place in it.

By Dirichlet's Theorem, the arithmetic progressions (arithmetic sequences) of terms $a + im$ and $(a + 2) + im$, $i = 0, 1, 2, 3, 4, \ldots$, contain infinitely many primes if a and $a + 2$ have no factors or divisors in common with m other than $+1$ and -1 (i.e., the integers, a and m, are relatively prime or coprime, as well as the integers, $a + 2$ and m). We now, e.g., let $a = 5$ and $m = 12$ (we could always assign values to a and m such that both a and $a + 2$ have no factors or divisors in common with m other than $+1$ and -1). Then we have the following sequences ($a + im$, $i = 0, 1, 2, 3, 4, \ldots$, and, $(a + 2) + im$, $i = 0, 1, 2, 3, 4, \ldots$):-

5, 17, 29, 41, 53, 65, 77, 89, 101, 113, 125, 137, 149, 161, 173, ...

7, 19, 31, 43, 55, 67, 79, 91, 103, 115, 127, 139, 151, 163, 175, ...

The integers in the above sequences which are underlined are prime numbers. The first sequence contains infinitely many primes. The second sequence contains infinitely many primes too. The corresponding integers in both the sequences, e.g., 5 and 7, 17 and 19, 29 and 31, and so on, differ from one another by 2, which is a characteristic of the twin primes, and, a number of twin primes is evident in both the sequences.

Since the twin primes (5 & 7, 17 & 19, 29 & 31, 41 & 43, 101 & 103, 137 & 139, and, 149 & 151) are evident in both the sequences, by the infinitude of the primes in both the sequences as postulated by Dirichlet's Theorem, and, the above lemma, which postulates that the elements of an infinite subset of an infinite set contain all the recursive significant properties of that set unless the process that selects the elements of the subset directly excludes a property, infinitely many of them must appear thereafter in both the sequences; so, there must be infinitely many twin primes, the twin primes representing a characteristic or feature of the prime numbers, the recurrent, identical mathematical structures postulated by the well-established self-similarity concept.

PART 2

Theorem:- There is an infinitude of twin primes.

Argument 1:-

Lemma: The development of the self-similarity concept has brought fame to Mitchell Feigenbaum, who has worked in the Los Alamos Laboratory in the early 1970s. The "distinguished presence" of the twin primes amongst the prime numbers, as described in the chapter, is analogous to the "self-similar mathematical pattern or structure" (which is the shape of the fig-tree itself) of the various parts of the fig-tree, i.e., its trunk to bough section, bough to branch section, branch to twig section and twig to twiglet section, in Feigenbaum's famous fig-tree example, and, such self-similar mathematical pattern or structure, or, fractal characteristic, could also be found in other aspects of nature, e.g., waves, turbulence or chaos, the structures of viruses and bacteria, polymers and ceramic materials, the universe and many others, even the movements of prices in financial markets, the growths of populations, the sound of music, the flow of blood through our circulatory system, the behaviour of people en masse, etc., which have all spawned a relatively new and important branch of mathematics with wide practical applications known as fractal geometry, which has been pioneered by Benoit Mandelbrot. As a matter of fact, self-similarity or fractal characteristic could be regarded as the fundamental mathematical aspect found in practically everything in nature including the numbers such as the prime numbers and twin primes which are the subjects of our investigation here, and, this new branch of mathematics, fractal geometry, besides having a great practical impact on us also gives us a deeper vision of the universe in which we live and our place in it.

We could generate an infinite list of primes, Q, by utilising Euclid's indirect (reductio ad absurdum) method of proving the infinity of the primes. In his proof, he first assumed that there is a finite list of prime numbers and then showed that there must exist an infinite number of additions to this list. The proof is as follows:-

Assume that n is the largest existing prime number in the following list of primes:

$$x = 2, 3, 5, 7, 11, 13, 17, 19, ..., n$$

There could be a prime, q, which is larger than n, computed as follows:

$$q = (2 \times 3 \times 5 \times 7 \times 11 \times 13 \times 17 \times 19 \times .... \times n) + 1$$

This number, q, could not be divided exactly by any of our primes (from 2 to n) and must be either a prime number itself which is larger than n or must be divisible by prime factors each of which is larger than n, both of which cases contradicting our original assumption that n is the largest existing prime, thus implying the infinity of the primes. In this manner we have added to our original list of primes, which is x. We could now repeat the process by including our new prime, q, or its prime factors if q is not prime in our list and generate a new number that is either a prime number that is larger than q or a composite number with prime factors that are each larger than q, i.e., we could generate some other new prime or primes that are not on our list of known primes. No matter how long our list of prime numbers is it is possible to find some new prime or primes. Thus, the list of primes is never-ending and infinite. In other words, Q is an infinite list of prime numbers.

We could in fact modify Euclid's indirect method slightly, but use the same reasoning, to generate another infinite list of prime numbers, P. (This method would have had served Euclid equally well in proving the infinity of the primes.) This argument takes the following form:-

Assume that n is the largest existing prime in the following list of prime numbers:

$$x = 2, 3, 5, 7, 11, 13, 17, 19, ..., n$$

There could be a prime number, p, which is larger than n, computed as follows:

$$p = (2 \times 3 \times 5 \times 7 \times 11 \times 13 \times 17 \times 19 \times .... \times n) - 1$$

(p is either a prime that is larger than n, or, it has prime factors each of which is larger than n; therefore, n could not be the largest existing prime, and, P is an infinite list of prime numbers.)

It is evident that Q - P = 2. In other words, $q_1 - p_1 = 2$, $q_2 - p_2 = 2$, $q_3 - p_3 = 2$, $q_4 - p_4 = 2$, and so on. This is the characteristic of twin primes.

We would here match P and Q for n = 2 to 19, which is as follows:-

n = 2 to 19	P	Q
1] $(2 \times 3) \pm 1$	5 = Prime	7 = Prime (Direct twin primes)
2] $(2 \times 3 \times 5) \pm 1$	29 = Prime	31 = Prime (Direct twin primes)
3] $(2 \times 3 \times 5 \times 7) \pm 1$	209 = 11 x 19 (Indirect twin primes: 11 & 13; 17 &19)	211 = Prime

4] $(2 \times 3 \times 5 \times 7 \times 11) \pm 1$ 2,309 = Prime   2,311 = Prime (Direct twin primes)

5] $(2 \times 3 \times 5 \times 7 \times 11 \times 13) \pm 1$ 30,029 = Prime   30,031 = 59 x 509
                                                      (Indirect twin primes: 59 & 61)

6] $(2 \times 3 \times 5 \times 7 \times 11 \times 13 \times 17) \pm 1$ 510,509 = 61 x 8,369   510,511 = 19 x 97 x 277
                               (Indirect twin primes: 59 & 61)   (Indirect twin primes: 17 & 19)

7] $(2 \times 3 \times 5 \times 7 \times 11 \times 13 \times 17 \times 19) \pm 1$ 9,699,689 =   9,699, 691 =
                                                  53 x 197 x 929   347 x 27,953
                                              (Indirect twin primes: (Indirect twin primes:
                                                  197 & 199)   347 & 349)

It is evident that P and Q, for n = 2 to 19, yield three pairs of direct twin primes and five pairs of different indirect twin primes which are the result of pairing the prime factors of the p's and the q's with the prime numbers which differ from each of them respectively by 2. Moreover, all these direct and indirect twin primes, a total of eight pairs of them, are found between n and Q and are of course larger than n, the assumed largest existing prime number; they are scattered amongst the twenty-five newly found prime numbers (which are all larger than n). (Note: The above computations are performed with an ordinary hand-held calculator which is capable of yielding results up to n = 19 only. For results for n > 19, a more powerful calculator or computer would be needed.)

Beside these direct and indirect twin primes between n and Q, more twin primes larger than n, and relatively plenty of them too, could be generated between n and Q by resorting to the Sieve of Eratosthenes - we shall call these twin primes the extraneous twin primes. For example, in (3) above, where n = 7 and Q = 211, 41 & 43, 71 &73, 101 & 103, etc., are extraneous twin primes which are all larger than n (n = 7) and n - 2 = 5, 5 & 7 here being twin primes, and, in (5) above, where n = 13 and Q = 30,031, 107 & 109, 137 &139, 227 & 229, 311 & 313, etc., are extraneous twin primes which are all larger than n (n = 13) and n - 2 = 11, 11 & 13 being twin primes. Hence, if we assume that any of these n - 2's and n's are the largest existing twin primes in the spirit of Euclid we would encounter contradictions from the extraneous twin primes (as well as the direct twin primes and/or the indirect twin primes) which are all larger (these being indirect arguments for the infinitude of the twin primes, which would be described below).

By induction:

P contains an infinite listing of direct primes, the p's, and, prime factors of the p's wherever the p's are non-prime, as proven above.

Q also contains an infinite listing of direct primes, the q's, and, prime factors of the q's wherever the q's are non-prime, as proven above.

Q - P = 2, which is a characteristic of the twin primes.

Both P and Q contain a number of direct twin primes (the result of the pairing of the direct primes in P and Q) and indirect twin primes (the result of the pairing of the prime factors of the p's and the q's in both P and Q with the prime numbers which differ from each of them respectively by 2) for n = 2 to 19, between n and Q, as is shown in the above example - in this case, a total of eight pairs of them scattered amongst the twenty-five newly found prime numbers, and everyone of these, twin primes and newly found prime numbers, is larger than n, the assumed largest existing prime, a contradiction. Beside these direct and indirect twin primes, more twin primes, the evidently more abundant extraneous twin primes between n and Q, are generated by the Sieve of Eratosthenes - these extraneous twin primes are all larger than n as well. All these are in line with the concept behind our arguments by contradiction, the indirect arguments - in these indirect arguments, contradiction implies an infinite number of primes, not excluding the twin primes, which are two primes that differ from one another by 2. (The indirect argument for the twin primes' infinitude is described below.) From the above-mentioned example of twin primes, and, many, many other examples of twin primes beside this, it is evident that twin primes represent a characteristic or feature of the prime numbers. This is in accordance with the above lemma, upon which the method of renormalization in perturbation theory is based, developed by Mitchell Feigenbaum in the 1970s, which postulates that there is a tendency of identical mathematical structures to recur on many levels. Within a given structure, there would be smaller copies of the same structure, their sizes being determined by the scaling factor. Feigenbaum found that at the utmost tips of the fig-tree, there is some mathematical structure which remains the same when its size is changed (enlarged) by a scaling factor of 4.669, which is found to be a constant like pi (3.142); this structure is the shape of the fig-tree itself; in other words, little whorls could be found within big whorls. Renormalization has been a well-established technique in chaos theory/fractal geometry and is a mathematical trick which functions rather like a microscope, zooming in on the self-similar structure, removing any approximations, and filtering out everything else. All this shows the universality of some features of chaos. That is, some kind of order or pattern could be found in or is inherent in disorder or chaos. Likewise, the "distinguished" appearances of the twin primes could be found amongst the chaotic (disorderly, random, patternless) array of infinite prime numbers, the twin primes being comparable to the little whorls found within the big whorls - they represent the recurrent, identical mathematical structures postulated by the above lemma.

Since by Euclid's proof the list of the primes is infinite, and, since the twin primes represent a characteristic or feature of the prime numbers as is described above, it could thus be concluded that there is an infinitude of twin primes. This is the result of the "reflection" principle. (For example, the characteristic of a mountain or infinite volume of sand is reflected in the characteristic of some grains of sand found there so that studying the characteristic of some grains of sand found there is sufficient for deducing the characteristic of the mountain or infinite volume of sand. The twin primes found in P and Q above could be taken to represent the "grains of sand" found in the "infinite volume of sand" which is represented by the infinite list of the primes. The "grains of sand" (which are in effect a subset) and the "infinite volume of sand" (which are in effect a set), here, share this same characteristic - the presence of twin primes within them.)

Thus, in general, some of the direct primes in P and Q would pair to form direct twin primes and some of the prime factors of the p's and the q's in both P and Q would pair with other prime numbers (which differ from each of them respectively by 2) to form indirect twin primes, not to mention the presence of the relatively abundant extraneous twin primes generated by the Sieve of Eratosthenes. All these results are representative of the afore-mentioned characteristic or "pattern" of the prime numbers, and, they are in accordance with the above lemma - they represent the recurrent, identical mathematical structures postulated by the above lemma.

By the above lemma and the "reflection" principle, the infinity of the primes in P (direct primes and prime factors) and the infinity of the primes in Q (direct primes and prime factors) imply that there is an infinitude of twin primes (direct twin primes, indirect twin primes and extraneous twin primes) generated by both P and Q together, and, the Sieve of Eratosthenes.

Hence, it is clear that the twin primes are infinite.

Argument 2:-

The following is indirect argument (argument by contradiction) of the infinity of the twin primes. As P and Q contain an infinitude of twin primes (direct and indirect), including the extraneous twin primes between n and Q generated by the Sieve of Eratosthenes, which are all larger than n, the assumed largest existing prime number, as proven above, this implies that if this n, and, n - 2, are both prime numbers (twin primes), they could not be the largest existing pair of twin primes. If we assume that these n and n - 2 are the largest existing twin primes, we would be contradicted by the above-mentioned argument. This is the indirect argument for the infinity of the twin primes. In the above-mentioned example, for n = 2 to 19, our respective assumed largest existing pairs of twin primes (n - 2 & n), 3 & 5, 5 & 7, 11 & 13, and, 17 & 19, are superceded (and "contradicted") by 29 & 31 (direct twin primes); 11 & 13, &, 17 & 19 (indirect twin primes); 59 & 61 (indirect twin primes); and, 197 & 199, &, 347 & 349 (indirect twin primes); respectively; not to mention the respective larger extraneous twin primes generated by the Sieve of Eratosthenes. In other words, it could be expected that all the direct twin primes and the indirect twin primes so produced (or generated), and the extraneous twin primes generated by the Sieve of Eratosthenes, as described above, would be larger than n (the assumed largest existing prime number), or, n - 2 and n (the assumed largest existing pair of twin primes) - this is incontrovertible indirect evidence of the infinitude of the primes and twin primes.

With the two formulae, (2 x 3 x 5 x 7 x 11 x 13 x 17 x 19 x ……. x n) - 1 and (2 x 3 x 5 x 7 x 11 x 13 x 17 x 19 x ……. x n) + 1 (and, the help of the Sieve of Eratosthenes), we could indeed generate two infinite lists of prime numbers, with an infinite number of twin primes (direct, indirect and extraneous) scattered here and there all over the place.

We state the following theorem:-

If we let n represent any successive prime number in a list of consecutive prime numbers which commences from the prime number 2 and assume it to be the largest existing prime number, then by the following statements:

p = (2 x 3 x 5 x 7 x 11 x 13 x 17 x 19 x .... n) - 1

q = (2 x 3 x 5 x 7 x 11 x 13 x 17 x 19 x .... n) + 1

since q - p = 2, if both p and q are prime numbers, in which case they are twin primes, they would be larger than n, and, if n - 2 is also a prime number, in which case these n and n - 2 are twin primes, and if we then assume that these n and n - 2 are the largest existing twin primes pair, we would find that these twin primes, p and q, would be larger than these n and n - 2, the assumed largest existing twin primes pair, hence contradicting the assumption that these n and n - 2 are the largest existing twin primes pair and proving the infinitude of the twin primes according to the principle of "reductio ad absurdum"; if, on the other hand, either of these p and q or both these p and q are not prime (i.e., they are composite numbers), then the prime factors of either of these p and q, or, both these p and q would be larger than these n and n - 2, and, if any of these prime factors pair with other prime numbers that differ from them by 2 then we would have a twin primes pair or pairs consisting of these prime factor or factors of p and/or q and another prime number or numbers that differ from these prime factor or factors by 2 which would be larger than these n and n - 2, the assumed largest existing twin primes pair, thereby contradicting this assumption and proving the infinity of the twin primes through the principle of "reductio ad absurdum"; this rule would always apply whenever p and q are both primes, i.e., they are twin primes, or, whenever p and/or q are non-prime, i.e., they are composite numbers, and their prime factor or factors pair with another prime number or numbers that differ from them by 2 to form a twin primes pair or pairs, when, at the same time, n and n - 2 are also prime numbers, i.e., twin primes. In other words: If p and q are both primes, i.e., twin primes, or, if a prime factor or factors of p and/or q pair with another prime number or numbers that differ from them by 2 to form a twin primes pair or pairs (when p and/or q are composite, i.e., non-prime), then n and n - 2 could not be the largest existing twin primes pair if it happened that n and n - 2 are both primes, i.e., twin primes.

Argument 3:-

The following describes two other ways of proving the infinity of the primes:-

(1)   Whatever number one picks, there is a prime larger than it.
(2)   Pick a number - N.
(3)   Multiply all the positive integers starting with 1 and ending with N.
(4)   I.e., form the factorial of N (written N!).
(5)   What one gets is divisible by every number up to N.
(6) When one adds 1 to N!, the result:

    (a) can't be a multiple of 2 (because it leaves 1 over, when you divide by 2).
    (b) can't be a multiple of 3 (because it leaves 1 over, when you divide by 3).
    (c) can't be a multiple of 4 (because it leaves 1 over, when you divide by 4).
    .
    .
    (n) can't be a multiple of N (because it leaves 1 over, when you divide by N).

(7)   In other words, N! + 1, if it is divisible at all (other than by 1 and itself) only

is divisible by numbers greater than N.

(8)   So, either N! + 1 is itself prime, or its prime divisors are greater than N.
(9)   In either case, there must exist a prime bigger than N.
(10)   The process holds no matter what number N is.
(11)   Whatever N is, there is a prime greater than N.
(12)   Thus, the infinitude of the primes. (Generalisation)

However, if we modify (6) above as follows the conclusion would still be the same:-

(6)   When one minuses 1 from N!, the result:

        (a)   can't be a multiple of 2 (because it leaves 1 over, when you divide by 2).
        (b)   can't be a multiple of 3 (because it leaves 2 over, when you divide by 3).
        (c)   can't be a multiple of 4 (because it leaves 3 over, when you divide by 4).
        .
        .
        .
        (n)   can't be a multiple of N (because it leaves N - 1 over, when you divide by N).

(7)   In other words, N! -1, if it is divisible at all (other than by 1 and itself) only is divisible by numbers greater than N.
(8)   So, either N! - 1 is itself prime, or its prime divisors are greater than N.
(9)   In either case, there must exist a prime bigger than N.
(10)   The process holds no matter what number N is.
(11)   Whatever N is, there is a prime greater than N.
(12)   Thus, the infinitude of the primes. (Generalisation)

Therefore: When N! - 1 and N! + 1 are both primes, or, when any of their prime divisors are partnered by primes which differ from them by 2 the result would be new twin primes pairs. By the infinitude of the primes generated by both N! + 1 and N! - 1 (which differ from one another by 2, a characteristic of the twin primes), as described above, and, the well-established self-similarity concept stated above, there would be an infinitude of such twin primes pairs, the twin primes being a characteristic or feature of the prime numbers, the recurrent, identical mathematical structures postulated by the self-similarity concept.

Argument 4:-

All odd numbers are either prime numbers or composite numbers (i.e., they contain prime factors), the prime numbers being the "atoms" or building-blocks of the odd numbers. Based on this premise, we carry on with the following steps:-

(1)   It is possible to generate new prime numbers. All we have to do is to first create a composite number by multiplying the consecutive odd numbers together, starting from the lowest odd number, 3 (which is a prime number), in consecutive, ascending order, and ending at any odd number (N), for example (a hand-held calculator yields results up to N = 17 only):

        (i) $3 \times 5 = 15$ (N = 5)
        (ii) $3 \times 5 \times 7 = 105$ (N = 7)
        (iii) $3 \times 5 \times 7 \times 9 = 945$ (N = 9)
        (iv) $3 \times 5 \times 7 \times 9 \times 11 = 10,395$ (N = 11)
        (v) $3 \times 5 \times 7 \times 9 \times 11 \times 13 = 135,135$ (N = 13)

(vi)   3 x 5 x 7 x 9 x 11 x 13 x 15 = 2,027,025 (N = 15)

(vii)   3 x 5 x 7 x 9 x 11 x 13 x 15 x 17 = 34,459,425 (N = 17)

.

.

.

(n)   3 x 5 x 7 x 9 x 11 x 13 x 15 x 17 x . . . . N

(2)   The results in (1) above would be composite numbers which are divisible by any of their odd factors (primes and multiples of primes), for example, 3 or 5 in (i), and, 3, 5, 7, 9, 11, 13, 15 or 17 in (vii).

(3)   Next, we attempt to make each of these results non-composite, i.e., turn each of these results into a prime number, by taking away 2 from each of them to get result X (which could turn out to be a prime number).

(4)   Then, we take away 2 from X, and, we could get another prime number, Y (which would of course be a twin prime if it is a prime number and if X is also a prime number).

(5)   For example:

(a)   In (1)(i) above, we would get the following twin primes:

      (3 x 5) - 2 = 13   (X)
      (3 x 5) - 2 - 2 = 11   (Y)

(b)   In (1)(ii) above, we would get the following twin primes:

      (3 x 5 x 7) - 2 = 103   (X)
      (3 x 5 x 7) - 2 - 2 = 101   (Y)

(c)   In (1)(iii) above, we would get the following composite number and prime:

      (3 x 5 x 7 x 9) - 2 = 943   (X) - Composite Number
      (3 x 5 x 7 x 9) - 2 - 2 = 941   (Y) - Prime Number

(d)   In (1)(iv) above, we would get the following composite number and prime:

      (3 x 5 x 7 x 9 x 11) - 2 = 10,393   (X) - Composite Number
      (3 x 5 x 7 x 9 x 11) - 2 - 2 = 10,391 (Y) - Prime Number

(e)   In (1)(v) above, we would get the following composite number and prime:

      (3 x 5 x 7 x 9 x 11 x 13) - 2 = 135,133   (X) - Composite Number
      (3 x 5 x 7 x 9 x 11 x 13) - 2 - 2 = 135,131 (Y) - Prime Number

(f)   In (1)(vi) above, we would get the following twin primes:

      (3 x 5 x 7 x 9 x 11 x 13 x 15) - 2 = 2,027,023   (X)
      (3 x 5 x 7 x 9 x 11 x 13 x 15) - 2 - 2 = 2,027,021 (Y)

(g)   In (1)(vii) above, we would get the following composite number and prime:

      (3 x 5 x 7 x 9 x 11 x 13 x 15 x 17) - 2 = 34,459,423   (X) - Prime Number
      (3 x 5 x 7 x 9 x 11 x 13 x 15 x 17) - 2 - 2 = 34,459,421 (Y) - Composite Number

.

.

.

(6)   X and Y would be larger than N and would never be divisible by any of the consecutive odd numbers, 3 to N. They are either prime numbers (which may pair to become twin primes) or composite numbers. The fact that X and Y are not divisible by any of the consecutive odd numbers, 3 to N, means that if they are each composite numbers their prime factors would be larger than N and if each of these prime factors pair with another prime to form a twin primes pair this twin pair would of course also be larger than N.

(7)   Prime numbers, and, twin primes are likely to appear in X and Y for the following reasons:

(a)   The lists of numbers to be multiplied together to create composite numbers in (1) above do not include even numbers and all the numbers are consecutive odd numbers (whereas Euclid's list of prime numbers in his indirect proof also includes the prime 2 which is an even number), leaving relatively little room or opportunity for composite numbers with prime factors larger than N to form after executing steps (3) and (4) above. The inclusion of even number(s) would cause an increase in the numerical value of the composite number consisting of odd factors, this composite number being only divisible by odd numbers, for example:

In the following list of numbers which includes even numbers: 2 x 3 x 4 x 5 x 6 x 7 x 8 x 9 x 10 x 11, the composite of odd numbers (3 x 5 x 7 x 9 x 11 = 10,395), 10,395, would be increased in numerical value by (2 x 4 x 6 x 8 x 10 = 3,840) 3,840 times to become (10,395 x 3,840 = 39,916,800) 39,916,800, due to the inclusion of the even numbers, 2, 4, 6, 8 and 10.
(If we, for example, attempt to convert the result of multiplying the composite of odd numbers by the even numbers, in this case 39,916,800, into twin primes by subtracting 1 from it (39,916,800 - 1 = 39,916,799 which is not a prime but a composite number) and then subtracting 2 from the last result (39,916,799 - 2 = 39,916,797 which is also a composite number), we are more likely to end up with composite numbers rather than primes or twin primes.)

(b)   X and Y are each smaller than the respective composites of the consecutive odd numbers, 3 to N (shown in (1) above), by 2 and 4 respectively, evidently leaving even less room or opportunity for composite numbers with prime factors larger than N to form after executing steps (3) and (4).

(c)   (5) above evidently shows that X and Y are likely to contain primes and twin primes.

(8)   Since X and Y if they are primes or if they pair to become twin primes, the prime factors of X and/or Y if X and/or Y are non-prime (composite) and any twin pairs formed by the prime factors of X and Y with other primes would be larger than N, this means that N would never be the largest existing prime number if it is prime and N - 2 and N would never be the largest existing twin primes pair if they are both primes. This is indirect evidence of the infinity of the primes and twin primes. In the indirect argument, the argument by contradiction, we assume that N, and, N - 2 and N are the largest existing prime and twin primes pair respectively. Then we find a larger prime and twin primes pair to contradict this assumption and prove the infinity of the primes and twin primes.

(9)   In the above-mentioned manner, we could go on creating new primes and twin primes ad infinitum.

(10) Thus, by (8) above and the well-established self-similarity concept stated above, we conclude the infinity of the primes and twin primes.

APPENDIX

The Sieve Of Eratosthenes

The Sieve of Eratosthenes is an algorithm for making tables of primes. The process of determining all the primes not greater than a number N is carried out by writing down all the numbers from 2 to N, removing those after 2 which are multiples of 2, those after 3 which are multiples of 3, and continuing until all multiples of primes not greater than the square root of N, except the primes themselves, have been removed. Only prime numbers would remain.

The following is an example of the generation of primes (and twin primes) by the Sieve of Eratosthenes:-

The following numbers from 2 to 50 are sieved by removing those which are multiples of 2, 3, 5 and 7 (square root of 50):

1 2 3 *4* 5 *6* 7 *8* *9* *10* 11 *12* 13 *14* *15* *16*

17 *18* 19 *20* *21* *22* 23 *24* *25* *26* *27* *28* 29 *30*

31 *32* *33* *34* *35* *36* 37 *38* *39* *40* 41 *42* 43 *44*

45 *46* 47 *48* *49* *50*

The bold, italised numbers have been sieved and removed.

The following prime numbers remain after sieving:

2, 3, 5, 7, 11, 13, 17, 19, 23, 29, 31, 37, 41, 43 and 47

The following are twin primes:

3 & 5, 5 & 7, 11 & 13, 17 & 19, 29 & 31, and, 41 & 43

CONCLUSION

A number of methods have been adopted in this chapter in solving the twin primes problem, which here lead us to conclude that the twin primes are infinite. Importantly, quite a number of ways of finding twin primes have been presented.

3 A SIMPLE SOLUTION FOR THE TWIN PRIMES CONJECTURE

This solution approaches the twin primes problem through the analysis of the intrinsic nature of the prime numbers.

Theorem:- The twin primes are infinite.

Argument:-
We note a very important intrinsic characteristic of the primes. Like all the houses in a neighbourhood or location which are separated from each other by the number of houses between them, the primes are also separated from each other by the number of integers separating them. The closest will of course be the prime neighbours separated by 2 integers (i.e., twin primes), followed next in proximity by the prime neighbours separated by 4 integers, then by the prime neighbours separated by 6 integers, the prime neighbours separated by 8 integers, the prime neighbours separated by 10 integers, the prime neighbours separated by 12 integers, and so on, by larger and larger intervals, to infinity, as is shown in the appendix. The twin primes are actually comparable to 2 closest neighbours living just next door to one another. There will always be 2 closest next-door neighbours, neighbours living 2 doors away, neighbours living 3 doors away, neighbours living 4 doors away, neighbours living 5 doors away, neighbours living 6 doors away, and so on, by greater and greater intervals, in any neighbourhood, any residential area; there will always be different intervals separating all the houses in a neighbourhood or location. Similarly, in the infinite list of the primes, there will always be different intervals separating all the primes, ranging from the smallest interval of 2 integers (in the case of the twin primes), 4 integers, 6 integers, 8 integers, 10 integers, 12 integers, and more and more integers, to an infinite number of integers, which is an intrinsic characteristic of the primes. In other words, there will always be intervals of various magnitudes or sizes (i.e., intervals of various numbers of integers) between, separating, all the primes in the infinite list of the primes, and, each of these intervals of various magnitudes or sizes should be intrinsically infinite in order that the list of the primes is infinite. The twin primes, which we are examining here, are not likely to be finite (as is evident from the appendix), and should of course be intrinsically infinite; in fact, to say that the twin primes are finite is like saying that next-door neighbours who are closest, in a neighbourhood or residential area, are rare and limited, which is absurd.

Hence, our conclusion that the twin primes are infinite.

APPENDIX

Anecdotal Evidence Of The Infinity Of The Twin Primes

TOP TWIN PRIMES IN 2000, 2001, 2007 & 2009
In the year 2000, $4648619711505 \times 2^{60000} \pm 1$ (18,075 digits) had been the top twin primes pair which had been discovered. In the year 2001, it only ranked eighth in the list of top 20 twin primes pairs, with $318032361 \cdot 2^{107001} \pm 1$ (32,220 digits) topping the list. In the year 2007, in the list of top 20 twin primes pairs, $318032361 \cdot 2^{107001} \pm 1$ (32,220 digits) ranked eighth, while $4648619711505 \times 2^{60000} \pm 1$ (18,075 digits) was nowhere to be seen; $2003663613*2^195000-1$ and $2003663613*2^195000+1$ (58,711 digits), which was discovered on January 15, 2007, by Eric Vautier (from France) of the Twin Prime Search (TPS) project in collaboration with PrimeGrid (BOINC platform), was at the top of the list. As at August 2009, $65516468355 \cdot 2^{333333}-1$ and $65516468355 \cdot 2^{333333}+1$ (100,355 digits) is at the top of the list of top 20 twin primes pairs, while $318032361 \cdot 2^{107001} \pm 1$ (32,220 digits) ranks 11th., and, $2003663613*2^195000-1$ and $2003663613*2^195000+1$ (58,711 digits) ranks second in this list.

We can expect larger twin primes than these extremely large twin primes, much larger ones, infinitely larger ones, to be discovered in due course.

<u>LIST OF PRIMES PAIRS FOR THE FIRST 2,500 CONSECUTIVE PRIMES, 2 TO 22,307, RANKED ACCORDING TO THEIR FREQUENCIES OF APPEARANCE</u>

S. No.	Ranking	Prime Pairs	No. Of Pairs	Percentage
(1)	1	primes pair separated by 6 integers	482	19.29 %
(2)	2	primes pair separated by 4 integers	378	15.13 %
(3)	3	primes pair separated by 2 integers (t. p.)	376	15.05 %
(4)	4	primes pair separated by 12 integers	267	10.68 %
(5)	5	primes pair separated by 10 integers	255	10.20 %
(6)	6	primes pair separated by 8 integers	229	9.16 %
(7)	7	primes pair separated by 14 integers	138	5.52 %
(8)	8	primes pair separated by 18 integers	111	4.44 %
(9)	9	primes pair separated by 16 integers	80	3.20 %
(10)	10	primes pair separated by 20 integers	47	1.88 %
(11)	11	primes pair separated by 22 integers	46	1.84 %
(12)	12	primes pair separated by 30 integers	24	0.96 %
(13)	13	primes pair separated by 28 integers	19	0.76 %
(14)	14	primes pair separated by 24 integers	16	0.64 %
(15)	15	primes pair separated by 26 integers	10	0.40 %
(16)	16	primes pair separated by 34 integers	9	0.36 %
(17)	17	primes pair separated by 36 integers	5	0.20 %
(18)	18	primes pair separated by 32 integers	2	0.08 %
(19)	18	primes pair separated by 40 integers	2	0.08 %
(20)	19	primes pair separated by 42 integers	1	0.04 %
(21)	19	primes pair separated by 52 integers	1	0.04 %

Total No. Of Primes Pairs In List: 2,498

It is evident in the above list that the primes pairs separated by 6 integers, 4 integers and 2 integers (twin primes), among the 21 classifications of primes pairs separated by from 2 integers to 52 integers (primes pairs separated by 38 integers, 44 integers, 46 integers, 48 integers & 50 integers are not among them, but, they are expected to appear further down in the infinite list of the primes), are the most dominant, important. There is a long list, an infinite list, of other primes pairs, besides those shown in the above list, which also play a part as the building-blocks of the infinite list of the integers.

The list of the integers is infinite. The list of the primes is also infinite. The infinite primes are the building-blocks of the infinite integers - the infinite odd integers are all either primes or composites of primes, and, the infinite even integers, except for 2 which is a prime, are all also composites of primes. Therefore, all the primes pairs separated by the integers of various magnitudes, as described above, can never all be finite. If there is any possibility at all for any of these primes pairs to be finite, there is only the possibility that a number of these primes pairs are finite (but never all of them). However, will it have to be the primes pairs separated by 2 integers or twin primes (which are the subject of our investigation here), which are the only primes pair, or, one among a number of primes pairs, which are finite? Why question only the infinity of the primes pairs separated by 2 integers, the twin primes? Are not the infinities of the primes pairs separated by 8 integers and more, whose frequencies of appearance are lower, as compared to those of the primes pairs which are separated by 6, 4 and 2 integers respectively, in the above list of primes pairs, more questionable? Why single out only the twin primes? (There are at least 18 other primes pairs, separated by from 8 integers to 52 integers, whose respective infinities should be more suspect, as is evident from the above list of primes pairs, if any infinities should be doubted. Evidently, the primes pairs separated by 2 integers (twin primes) are not that likely to be finite.)

4 ANOTHER SIMPLE SOLUTION FOR THE TWIN PRIMES CONJECTURE

This solution approaches the twin primes problem through the analysis of the composite numbers.

Theorem:- The twin primes are infinite.

Argument:-
The integers or whole numbers comprise of both the even and odd integers, and, both the prime numbers and composite numbers, which are all infinite. (The primes had been proven to be infinite by Euclid long ago, and the composites of primes are also infinite, which is implied of course by the infinitude of the primes.) The even integers after 2 are composites of even, or, both even and odd, primes (there is only 1 even prime, i.e., 2, and it is always present in the composites which make up the even integers), e.g., $4 = 2 \times 2$, $8 = 2 \times 2 \times 2$, $12 = 2 \times 2 \times 3$, $30 = 2 \times 3 \times 5$, $32 = 2 \times 2 \times 2 \times 2 \times 2$, $36 = 2 \times 2 \times 3 \times 3$, etc., while the odd integers are comprised of both the primes, e.g., 3, 5, 7, 11, 13, 17, 19, etc., and composites of odd primes, e.g., $9 = 3 \times 3$, $15 = 3 \times 5$, $21 = 3 \times 7$, $27 = 3 \times 3 \times 3$, $35 = 5 \times 7$, $57 = 3 \times 19$, etc.

All the primes in the infinite list of primes are separated by integers ranging from 2 (in the case of the twin primes), 4, 6, 8, 10, 12, and upwards, to infinity, as is shown in the appendix. The question is whether the twin primes, i.e., the primes pairs separated by 2 integers, are infinite. However, as is evident from the appendix, the list of the twin primes is not likely to be finite and can be expected to be infinite. We will proceed to prove this.

The Prime Number Theorem, which describes the distribution of the primes, is familiar. The primes have been known as the basic units, atoms or building-blocks of the integers, both even and odd ones. We will show the indispensable role of the twin primes, i.e., the primes pairs separated by 2 integers, as well as the primes pairs separated by 4 integers, 6 integers, 8 integers, 10 integers, 12 integers, and more, to infinity, and the other primes, in the formation or construction of the composite numbers. We will first of all do so by analysing the construction or composition of the composite numbers, both even and odd ones - we will do so by deconstructing a number of composite numbers. A Composite Number Theorem which describes the distribution of the composite numbers, as well as their building-blocks, the prime numbers, is a possibility. We note that each composite number, whether even or odd, is the unique product of a particular, unique set of prime numbers only, and, no product of a different set of primes can ever possibly result in the same composite number. We demonstrate this fact through some examples, which are as follows:-

Even Composite Numbers
(i) $10 = 2 \times 5$ (only)
(ii) $34 = 2 \times 17$ (only)
(iii) $106 = 2 \times 53$ (only)
(iv) $258 = 2 \times 3 \times 43$ (only)
(v)   Etc. to infinity

Odd Composite Numbers
(i) $21 = 3 \times 7$ (only)
(ii) $99 = 3 \times 3 \times 11$ (only)
(iii) $325 = 5 \times 5 \times 13$ (only)
(iv) $451 = 11 \times 41$ (only)
(v)   Etc. to infinity

It is clear from these examples that no products of other primes can ever possibly produce the above-mentioned composites besides the products of the primes shown above, which are unique.

Lemma:

This is in accordance with the Fundamental Theorem of Arithmetic or Unique Factorisation Theorem, which states that there is only one possible combination of primes which will multiply together to produce any particular number, e.g., the only combination of primes which will produce the number 2,079 is: 3 x 3 x 3 x 7 x 11.

In the same manner, the following numbers are also uniquely factorised:

    63 = 3 x 3 x 7 (only)
    153 = 3 x 3 x 17 (only)
    1,021,020 = 2 x 2 x 3 x 5 x 7 x 11 x 13 x 17 (only)

In other words, every positive whole number which is not prime can be broken up into prime factors, and, this can happen in only one way.

By this lemma, we can produce an infinitude of unique (only ones possible) composite numbers by "playing around" with or manipulating any list of prime numbers. For example, with the following list of primes:

    2, 3, 5, 7, 11, 13, 17, 19

we obtain the following unique (only ones possible) composites:

    39 = 3 x 13 (only)
    78 = 2 x 3 x 13 (only)
    182 = 2 x 7 x 13 (only)
    2,261 = 7 x 17 x 19 (only)
    3,230 = 2 x 5 x 17 x 19 (only)
    46,189 = 11 x 13 x 17 x 19 (only)
    62,985 = 3 x 5 x 13 x 17 x 19 (only)
    746,130 = 2 x 3 x 5 x 7 x 11 x 17 x 19 (only)
    Etc.

Similarly, with the following twin primes (in bold), and, other primes, we obtain the following unique (only ones possible) composites:

    15 = **3** x **5** (only)
    35 = **5** x **7** (only)
    70 = 2 x **5** x **7** (only)
    143 = **11** x **13** (only)
    286 = 2 x **11** x **13** (only)
    323 = **17** x **19** (only)
    104,329 = **17** x **19** x **17** x **19** (only)
    344,285 = 3 x 11 x **17** x **19** x **17** x **19** (only)
    46,189 = **11** x **13** x **17** x **19** (only)   (13 & 17 are a primes pair separated by 4 integers)
    3,417,986 = 2 x **11** x **13** x **17** x **19** x 37 (only)   (13 & 17 are a primes pair separated by 4 integers)
    10,403 = **101** x **103** (only)
    11,663 = **107** x **109** (only)
    22,499 = **149** x **151** (only)
    121,103 = **347** x **349** (only)
    435 = 3 x **5** x 29 (only)
    34,320 = 2 x 2 x 2 x 2 x 3 x **5** x **11** x **13** (only)

62,418 = 2 x 3 x **101** x **103** (only)
12,621,939 = 3 x 11 x 17 x **149** x **151** (only)
1,616,615 = **5** x **7** x **11** x **13** x **17** x **19** (only)   (7 & 11 and 13 & 17 are each a primes pair separated by 4 integers)
74,364,290 = 2 x **5** x **7** x **11** x **13** x **17** x **19** x 23 (only)   (7 & 11 and 13 & 17 are each a primes pair separated by 4
419,868 = 2 x 2 x 3 x 3 x **107** x **109** (only)   integers)
1,331,584 = 2 x 2 x 2 x 2 x 2 x 2 x 2 x **101** x **103** (only)
1,001,623 = 7 x **17** x **19** x 443 (only)
Etc. to infinity

The following is the product of primes for composite numbers in abbreviated form:-

$$c = \prod_{p\ prime} p$$

In the formation or construction of these unique (only ones possible) composites described above, there will be an infinitude of combinations of the twin primes pairs, which we are here examining, with the other primes in the infinite list of the primes. As is shown above, a unique (only one possible) composite can be the product of a twin primes pair (e.g., 3 x 5). It can be the product of a twin primes pair with another prime or primes (e.g., 3 x 5 x 19 x 79). It can be the product of a twin primes pair with itself (e.g., 3 x 5 x 3 x 5). It can be the product of a twin primes pair with itself and another prime or primes (e.g., 3 x 5 x 3 x 5 x 23 x 89). It can be the product of a twin primes pair with another twin primes pair or other twin primes pairs (e.g., 11 x 13 x 227 x 229 x 461 x 463) It can be the product of a twin primes pair with another twin primes pair or other twin primes pairs and another prime or primes (e.g., 11 x 13 x 227 x 229 x 461 x 463 x 701 x 1,459 x 2,447). It is evident that due to the infinitude of the primes, the amount of such combinations will be infinite.

The reasoning to be brought up hereafter is subtle. There should be an infinitude of products of primes pairs separated by 2 integers (i.e., twin primes, e.g., 29 & 31 and 41 & 43), prime pairs separated by 4 integers (e.g., 7 & 11 and 13 & 17), prime pairs separated by 6 integers (e.g., 1,117 & 1,123 and 1,861 & 1,867), primes pairs separated by 8 integers (e.g., 2,459 & 2,467 and 4,289 & 4,297), primes pairs separated by 10 integers (e.g., 6,691 & 6,701 and 10,321 & 10,331), primes pairs separated by 12 integers (e.g., 9,649 & 9,661 and 11,399 & 11,411), and so on, to infinity, and other primes from the infinite list of the primes, involved in the formation or construction of the composite numbers, since all these primes pairs separated by intervals of various magnitudes or sizes (from 2 integers in the case of the twin primes, 4 integers, 6 integers, 8 integers, 10 integers, 12 integers, and so on, upwards to infinity), and the other primes from the infinite list of the primes are evidently the building-blocks of different configurations which are necessary for the construction of the infinite unique (only ones possible) composites through the unique (only one possible) product of primes pairs separated by intervals of various magnitudes or sizes and other primes as well, all these primes being the unique (only ones possible) factors of the unique (only ones possible) composites, in line with the above lemma. The formation or construction of such unique (only ones possible) composites, which are infinite, can be described, e.g., as follows:-

$$c = p \times p_1 \times p_2 \times p_3 \times p_4 \times \ldots\ldots \quad (p_1\ \&\ p_2,\ p_3\ \&\ p_4\ being\ twin\ primes\ pairs)$$

It can thus be stated that the infinite list of the composite numbers comprises of the respective infinite lists of products of primes pairs separated by 2 integers (twin primes), primes pairs separated by 4 integers, primes pairs separated by 6 integers, primes pairs separated by 8 integers, primes pairs separated by 10 integers, primes pairs separated by 12 integers, and so on, upwards to infinity, and other primes.

All the respective lists of products relating to the primes pairs separated by intervals of various magnitudes or sizes and other primes which produce composites, which have been described above, are unique, the only ones of their kind possible, obtainable only through unique factorisation, which cannot be replaced or substituted through multiplying other prime numbers or factors (to produce the same composite number), the primes pairs separated by intervals of various magnitudes or sizes and the other primes being the building-blocks of different configurations which are necessary for the construction of the infinite unique (only ones possible) composites through the unique (only one possible) product of primes pairs separated by intervals of various magnitudes or sizes and other primes as well, which implies that an infinitude of them, since they are the building-blocks of different configurations, needs to exist in order that the list of the composites, of which they are the building-blocks of different configurations, is infinite, *more so* since they are unique (only ones possible) and cannot be replaced or substituted through multiplying other prime numbers or factors (to produce the same composite number). Since unique factorisation means that the products of all these primes pairs separated by intervals of various magnitudes or sizes and other primes as well will only produce unique (only ones possible) composites, in order for there to be larger and larger unique (only ones possible) composites to be produced we will need to get products of larger and larger primes pairs (i.e., primes pairs separated by 2 integers (twin primes), primes pairs separated by 4 integers, primes pairs separated by 6 integers, primes pairs separated by 8 integers, primes pairs separated by 10 integers, primes pairs separated by 12 integers, and so on, upwards to infinity) and larger and larger primes. To cater for the production of these larger and larger unique (only ones possible) composites to infinity, the supply of the primes pairs separated by 2 integers (twin primes), primes pairs separated by 4 integers, primes pairs separated by 6 integers, primes pairs separated by 8 integers, primes pairs separated by 10 integers, primes pairs separated by 12 integers, and so on, all the way upwards, and, the other primes, has also to be infinite, e.g., certain larger and larger unique (only ones possible) composites must necessarily have larger primes pairs separated by 2 integers (twin primes), larger primes pairs separated by 4 integers, larger primes pairs separated by 6 integers, larger primes pairs separated by 8 integers, larger primes pairs separated by 10 integers, larger primes pairs separated by 12 integers, and so on, upwards to infinity, and, other larger primes, as their *unique prime factors*, or, *only possible prime factors*, which implies that all these primes pairs separated by intervals of various magnitudes or sizes and the other primes need to be infinite in order to cater for the production of the infinite list of the larger and larger unique (only ones possible) composites. The infinite list of integers or whole numbers comprises of the infinite list of the prime numbers and the infinite list of the composite numbers, as is described above, which means that all these building-blocks of different configurations are also the building-blocks of the infinite integers or whole numbers.

What forms will these building-blocks of different configurations, these primes pairs separated by intervals of various magnitudes or sizes and the other primes, these *unique prime factors*, take, in the formation or construction of the unique (only ones possible) composites? Any primes pair in the group can be the product of one another. The primes pair can also be the product of itself with another prime or primes. It can also be the product of itself with itself. It can be the product of itself with itself and another prime or primes too. It can be the product of itself with another primes pair or other primes pairs as well. It can also be the product of itself with another primes pair or other primes pairs and another prime or primes. It is clear that due to the infinitude of the primes the amount of such combinations will be infinite. In this manner, an infinite quantity of larger and larger unique (only ones possible) composites can be produced, evidently owing to the fact that the primes involved are *unique prime factors* and larger primes. The product of a larger primes pair will always produce a larger composite number than the product of a smaller primes pair, e.g., the product of the larger twin primes pair 17 x 19 produces the larger unique (only one possible) composite number 323, while the product of the smaller twin primes pair 11 x 13 produces the smaller unique (only one possible) composite number 143. This fact implies that since the composites are infinite, the primes pairs separated by intervals of various magnitudes or sizes and the other primes, which are the composites' *unique prime factors*, will also be infinite, the infinite list of these primes pairs separated by intervals of various magnitudes and sizes and the other primes providing larger and larger prime factors for producing the larger and larger unique (only ones possible) composites. If we liken the infinite list of the integers or whole numbers to an infinitely large jigsaw puzzle, there will be an infinitude of "gaps of certain sizes", or, composites of certain values, that can only be filled up by the products of certain unique (only ones possible) primes from among the primes pairs separated by 2 integers (twin primes), primes pairs separated by 4 integers, primes pairs separated by 6 integers, primes pairs separated by 8 integers, primes pairs separated by 10 integers, primes pairs separated by 12 integers, and so on, upwards to infinity, and the other primes - these unique (only ones possible) primes are the *unique prime factors* of the unique (only ones possible) composites so constructed by the products of these unique (only ones possible) primes, as per the above lemma. This necessitates the infinitude of these unique (only ones possible) primes, i.e., the primes pairs separated by 2 integers (twin primes), primes pairs separated by 4 integers, primes pairs separated by 6 integers, primes pairs separated by 8 integers, primes pairs separated by 10 integers, primes pairs separated by 12 integers, and so on, upwards to infinity, and the other primes, for the purpose of filling up the "gaps of certain sizes", or, composites of certain values, which are infinite.

We bring up the following example of 10 larger and larger unique (only ones possible) composites which are the results of the products of 10 consecutive larger and larger twin primes pairs to show the "unique factorisation" effect:-

    (i)   15 = 3 x 5 (only)
    (ii)   35 = 5 x 7 (only)
    (iii) 143 = 11 x 13 (only)

(iv)   323 = 17 x 19 (only)
(v)   899 = 29 x 31 (only)
(vi)   1,763 = 41 x 43 (only)
(vii)   3,599 = 59 x 61 (only)
(viii)   5,183 = 71 x 73 (only)
(ix)   10,403 = 101 x 103 (only)
(x)   11,663 = 107 x 109 (only)

.
.
.

All the primes, or, building-blocks of different configurations, being *unique prime factors* of the unique (only ones possible) composites, are a necessity, a must-have, essential, compulsory, and, indispensable, for the construction of the latter, since there can never be any replacements or substitutes for them. The larger the unique (only ones possible) composites are, the larger their prime factors (i.e., these mentioned *unique prime factors*) therefore have to be. Since the unique (only ones possible) composites increase to infinity, it implies that these *unique prime factors* (which include the twin primes that are the subject of our examination), which are a necessity and are not substitutable, also have to increase to infinity. In the case of the twin primes, as in the case of the other primes in the infinite list of the primes, larger and larger twin primes, as well as larger and larger primes pairs separated by larger intervals than the twin primes' and other larger and larger primes, will be needed here and there for the production of larger and larger unique (only ones possible) composites. Thus, the twin primes, being *unique prime factors* of the infinite unique (only ones possible) composites, together with the other primes pairs separated by larger intervals and the other primes, are necessarily infinite.

In summing up, we state that since the twin primes, together with the other primes pairs separated by larger intervals and the other primes, are not replaceable or substitutable by other primes, and are a necessity, a must-have, essential, compulsory, and, indispensable, for the construction of the unique (only ones possible) composites, being the latter's atoms or building-blocks of different configurations, the latter's *unique prime factors* as per the above lemma, and since these unique (only ones possible) composites are infinite, it is implied that the twin primes, the other primes pairs separated by larger intervals and the other primes, being the latter's atoms or building-blocks of different configurations, also have to be infinite, larger and larger twin primes, primes pairs separated by larger intervals, and other primes, being essential, compulsory and necessary for the formation or construction of the larger and larger unique (only ones possible) composites and not replaceable or substitutable, becoming in effect the *unique prime factors*, or, *only possible prime factors*, of the latter, *more so* that they have to be around all the time, to be infinite, since they are essential, compulsory, necessary and cannot be replaced or substituted by other primes (to produce the same composite number) besides being the atoms or building-blocks of different configurations of the unique (only ones possible) composites.

APPENDIX

Anecdotal Evidence Of The Infinity Of The Twin Primes

TOP TWIN PRIMES IN 2000, 2001, 2007 & 2009

In the year 2000, 4648619711505 x $2^{60000} \pm 1$ (18,075 digits) had been the top twin primes pair which had been discovered. In the year 2001, it only ranked eighth in the list of top 20 twin primes pairs, with $318032361 \cdot 2^{107001} \pm 1$ (32,220 digits) topping the list. In the year 2007, in the list of top 20 twin primes pairs, $318032361 \cdot 2^{107001} \pm 1$ (32,220 digits) ranked eighth, while 4648619711505 x $2^{60000} \pm 1$ (18,075 digits) was nowhere to be seen; 2003663613*2^195000-1 and 2003663613*2^195000+1 (58,711 digits), which was discovered on January 15, 2007, by Eric Vautier (from France) of the Twin Prime Search (TPS) project in collaboration with PrimeGrid (BOINC platform), was at the top of the list. As at August 2009, $65516468355 \cdot 2^{333333}-1$ and $65516468355 \cdot 2^{333333}+1$ (100,355 digits) is at the top of the list of top 20 twin primes pairs, while $318032361 \cdot 2^{107001} \pm 1$ (32,220 digits) ranks 11th, and, 2003663613*2^195000-1 and 2003663613*2^195000+1 (58,711 digits) ranks second in this list.

We can expect larger twin primes than these extremely large twin primes, much larger ones, infinitely larger ones, to be discovered in due course.

LIST OF PRIMES PAIRS FOR THE FIRST 2,500 CONSECUTIVE PRIMES, 2 TO 22,307, RANKED ACCORDING TO THEIR FREQUENCIES OF APPEARANCE

S. No.	Ranking	Prime Pairs	No. Of Pairs	Percentage
(1)	1	primes pair separated by 6 integers	482	19.29 %
(2)	2	primes pair separated by 4 integers	378	15.13 %
(3)	3	primes pair separated by 2 integers (t. p.)	376	15.05 %
(4)	4	primes pair separated by 12 integers	267	10.68 %
(5)	5	primes pair separated by 10 integers	255	10.20 %
(6)	6	primes pair separated by 8 integers	229	9.16 %
(7)	7	primes pair separated by 14 integers	138	5.52 %
(8)	8	primes pair separated by 18 integers	111	4.44 %
(9)	9	primes pair separated by 16 integers	80	3.20 %
(10)	10	primes pair separated by 20 integers	47	1.88 %
(11)	11	primes pair separated by 22 integers	46	1.84 %
(12)	12	primes pair separated by 30 integers	24	0.96 %
(13)	13	primes pair separated by 28 integers	19	0.76 %
(14)	14	primes pair separated by 24 integers	16	0.64 %
(15)	15	primes pair separated by 26 integers	10	0.40 %
(16)	16	primes pair separated by 34 integers	9	0.36 %
(17)	17	primes pair separated by 36 integers	5	0.20 %
(18)	18	primes pair separated by 32 integers	2	0.08 %
(19)	18	primes pair separated by 40 integers	2	0.08 %
(20)	19	primes pair separated by 42 integers	1	0.04 %
(21)	19	primes pair separated by 52 integers	1	0.04 %

Total No. Of Primes Pairs In List: 2,498

It is evident in the above list that the primes pairs separated by 6 integers, 4 integers and 2 integers (twin primes), among the 21 classifications of primes pairs separated by from 2 integers to 52 integers (primes pairs separated by 38 integers, 44 integers, 46 integers, 48 integers & 50 integers are not among them, but, they are expected to appear further down in the infinite list of the primes), are the most dominant, important. There is a long list, an infinite list, of other primes pairs, besides those shown in the above list, which also play a part as the building-blocks of the infinite list of the integers.

The list of the integers is infinite. The list of the primes is also infinite. The infinite primes are the building-blocks of the infinite integers - the infinite odd integers are all either primes or composites of primes, and, the infinite even integers, except for 2 which is a prime, are all also composites of primes. Therefore, all the primes pairs separated by the integers of various magnitudes, as described above, can never all be finite. If there is any possibility at all for any of these primes pairs to be finite, there is only the possibility that a number of these primes pairs are finite (but never all of them). However, will it have to be the primes pairs separated by 2 integers or twin primes (which are the subject of our investigation here), which are the only primes pair, or, one among a number of primes pairs, which are finite? Why question only the infinity of the primes pairs separated by 2 integers, the twin primes? Are not the infinities of the primes pairs separated by 8 integers and more, whose frequencies of appearance are lower, as compared to those of the primes pairs which are separated by 6, 4 and 2 integers respectively, in the above list of primes pairs, more questionable? Why single out only the twin primes? (There are at least 18 other primes pairs, separated by from 8 integers to 52 integers, whose respective infinities should be more suspect, as is evident from the above list of primes pairs, if any infinities should be doubted. Evidently, the primes pairs separated by 2 integers (twin primes) are not that likely to be finite.)

5 ANOTHER SOLUTION FOR THE TWIN PRIMES CONJECTURE

INTRODUCTION

Many believe that the twin primes are infinite. In fact, twin primes pairs could easily be found among the integers. There is evidently no region of the natural number system so remote that it lies beyond the largest twin primes pair. It is even possible to forecast the approximate number of twin primes pairs found in any region of the natural number system.

The occurrence of twin primes pairs is evidently unpredictable or random. This means that the chance of 2 numbers x and x + 2 being prime (twin primes) is somewhat similar to the chance of getting heads on 2 successive tosses of a coin. If 2 successive tosses of a coin are independent, the chance of success of obtaining heads for the 2 successive tosses of the coin is the product of the chances of success of obtaining a head for each toss of the coin. As each coin has probability ½ of coming up heads with a toss, 2 coins would have probability ½ x ½ = ¼ of coming up a pair of heads with a toss.

The prime number theorem, which had been proven, states that if n is a large number, and we select a number x at random between 0 and n, the chance that x is prime would be approximately $1/\log n$, the larger n is, the better would be the approximation given by $1/\log n$ to the proportion of primes in the numbers up to n. Like 2 coins coming up heads, the chance that both x and x + 2 are prime (twin primes) would be approximately $1/(\log n)^2$. That is, there would be approximately $n/(\log n)^2$ twin primes pairs between 0 and n. As n goes to infinity, this fraction approaches infinity. This represents a quantitative version of the twin primes conjecture.

As x + 2 being prime depends on the fact that x is already prime, we should modify the estimate $n/(\log n)^2$ to $(1.32032..)n/(\log n)^2$.

The following is a comparison between the twin primes predicted by the above formula and the twin primes found, where the agreement is evidently very good:-

INTERVAL	TWIN PRIMES	
	PREDICTED	FOUND
100,000,000 - 100,150,000	584	601
1,000,000,000 - 1,000,150,000	461	466
10,000,000,000 - 10,000,150,000	374	389
100,000,000,000 - 100,000,150,000	309	276
1,000,000,000,000 - 1,000,000,150,000	259	276
10,000,000,000,000 - 10,000,000,150,000	221	208
100,000,000,000,000 - 100,000,000,150,000	191	186
1,000,000,000,000,000 - 1,000,000,000,150,000	166	161

All this represents numerical evidence that the twin primes are infinite as we could find more twin primes pairs whenever we look for them. But the proof is lacking.

Theorem:- The list of the twin primes pairs is infinite.

Argument:-

Lemma 1: According to the precepts of fractal geometry and group theory, symmetry is a very important, intrinsic part of nature. There is symmetry all around us and within us. There is evident symmetry in human bodies, the structures of viruses and bacteria, polymers and ceramic materials, the permutations of numbers, the universe and many others, even the movements of prices in financial markets, the growths of populations, the sound of music, the flow of blood through our circulatory system, the behavior of people en masse, etc. In other words, regularity, pattern, order, uniformity or symmetry is evident everywhere.

The reasoning here makes use of a very important idea in fractal geometry and group theory, namely, symmetry.

A prime number is an integer which is divisible only by 1 and itself, e.g., 2, 3, 7, 19, etc. A twin primes pair are 2 primes which differ from one another by 2, e.g., 5 & 7, 11 & 13, 17 & 19, and, 29 & 31, etc. A composite number or non-prime is a product of primes or prime factors, e.g., the composite numbers 15 is the product of 2 primes, 3 and 5 (15 = 3 x 5), and 231 is the product of 3 primes, 3, 7 and 11 (231 = 3 x 7 x 11), etc. The integers or whole numbers are either primes or composites and are infinite.

The primes, which Euclid had proven to be infinite, are the atoms or building-blocks of the infinite integers or whole numbers, which comprise of the infinite list of the odd numbers that are all either primes or products of primes (i.e., composites), and, the infinite list of the even numbers that are all products of primes (i.e., composites, with the exception of 2 which is a prime, e.g., 6 = 2 x 3, 8 = 2 x 2 x 2 and 10 = 2 x 5, etc.). The infinite list of the integers or whole numbers may be classified as an infinite group, with various symmetries, subgroups and infinite elements, hidden within it. These various symmetries, subgroups and infinite elements, within this infinite group may be classified as follows:-

(1) Subgroup A: Infinite consecutive primes such as 2, 3, 5, 7, 11, 13, 17, 19, 23, 29, 31, etc. to infinity .... , separated by 2 integers (twin primes), 4 integers, 6 integers, 8 integers, 10 integers, etc. to infinity .... , which, incidentally, except for 2, are all odd numbers; this splitting up of the subgroup into infinite elements is shown below:
  (i)   Element A1: Infinite list of all the primes pairs separated by 2 integers (twin primes) .... (Example: 17 & 19)
  (ii) Element A2: Infinite list of all the primes pairs separated by 4 integers/1 odd composite - single composite .... (Example: 79 & 83 separated by 81)
  (iii) Element A3: Infinite list of all the primes pairs separated by 6 integers/2 consecutive odd composites - twin composites .... (Example: 47 & 53

separated by 49 & 51)

(iv) <u>Element A4</u>: Infinite list of all the primes pairs separated by 8 integers/3 consecutive odd composites - "triple" composites .... (Example: 359 & 367 separated by 361, 363 & 365)

(v) <u>Element A5</u>: Infinite list of all the primes pairs separated by 10 integers/4 consecutive odd composites - "four-ple" composites .... (Example: 709 & 719 separated by 711, 713, 715 & 717)

.

.

.

(2) <u>Subgroup B</u>: Infinite consecutive odd composites such as 9, 15, 21, 25, 27, 33, 35, 39, 45, 49, etc. to infinity ...., of "infinite sizes" sandwiched between 2 primes; this splitting up of the subgroup into infinite elements is shown below:

(i) <u>Element B1</u>: Infinite list of all "1 odd composite sandwiched between 2 primes - single composite" .... (Example: 9 sandwiched between the primes 7 & 11)

(ii) <u>Element B2</u>: Infinite list of all "2 consecutive odd composites sandwiched between 2 primes - twin composites" .... (Example: 253 & 255 sandwiched between the primes 251 & 257)

(iii) <u>Element B3</u>: Infinite list of all "3 consecutive odd composites sandwiched between 2 primes - "triple" composites" .... (Example: 685, 687 & 689 sandwiched between the primes 683 & 691)

(iv) <u>Element B4</u>: Infinite list of all "4 consecutive odd composites sandwiched between 2 primes - "four-ple" composites" .... (Example: 2,769, 2,771, 2,773 & 2,775 sandwiched between the primes 2,767 & 2,777)

(v) <u>Element B5</u>: Infinite list of all "5 consecutive odd composites sandwiched between 2 primes - "five-ple" composites" .... (Example: 19,291, 19,293, 19,295, 19,297 & 19,299 sandwiched between the primes 19,289 & 19,301)

.

.

.

(3) <u>Subgroup C</u>: Infinite consecutive odd composites separated by 4 integers and 6 integers respectively; this splitting up of the subgroup into the 2 infinite elements is shown below:

   (i) <u>Element C1</u>: Infinite list of all "2 consecutive odd composites separated by 4 integers/1 prime" .... (Example: 209 & 213 separated by the prime 211)

   (ii) <u>Element C2</u>: Infinite list of all "2 consecutive odd composites separated by 6 integers/2 primes" .... (Example: 279 & 285 separated by the twin primes 281 & 283)

(4) <u>Subgroup D</u>: Infinite single primes and twin primes separating 2 consecutive odd composites; this splitting up of the subgroup into the 2 infinite elements is shown below:

   (i) <u>Element D1</u>: Infinite list of all the single primes separating 2 consecutive odd composites .... (Example: 23 separating the 2 consecutive odd composites 21 & 25)

   (ii) <u>Element D2</u>: Infinite list of all the twin primes separating 2 consecutive odd composites .... (Example: 11 & 13 separating the 2 consecutive odd composites 9 & 15)

(5) <u>Subgroup E</u>: Infinite consecutive even composites such as 4, 6, 8, 10, 12, 14, 16, 18, 20, 22, etc. to infinity ...., all separated by only 2 integers; this subgroup may be classified as a single infinite element .... There is always 1 even number between a twin primes pair, which is separated by 2 integers, a prime and a composite which are separated by 2 integers, and, 2 composites which are separated by 2 integers. That is, the even numbers are always found in Subgroup A, Subgroup B, Subgroup C and Subgroup D above always evenly spaced out in consecutive order by 2 integers.

There is an evident symmetry in the above-mentioned infinite group, which would be broken if any of the elements within were to be finite. There are close interlinks between all the various infinite elements in all the five subgroups above, e.g., the infinity of the list of all the primes pairs each separated by 6 integers (Element A3/Subgroup A) implies the infinity of the list of all the "2 consecutive odd composites sandwiched between 2 primes - twin composites" (Element B2/Subgroup B) and vice versa, the infinity of the list of all the primes pairs each separated by 2 integers (twin primes) (Element A1/SubgroupA) implies the infinity of the list of all the "2 consecutive odd composites separated by 6 integers/2 primes" (Element C2/Subgroup C) and vice versa, the infinity of the list of all the infinite elements (A1, A2, A3, etc. to infinity) in Subgroup A above, which represents the infinity of the list of the primes which Euclid had in fact proven, implies the infinity of the list of all the "2 consecutive odd composites separated by 4 integers/1 prime" (Element C1/Subgroup C) and vice versa, the infinity of the list of all the "2 consecutive odd composites separated by 6 integers/2 primes" (Element C2/Subgroup C) implies the infinity of the list of all the twin primes separating 2 consecutive odd composites (Element D2/Subgroup D) and vice versa, the infinity of all the lists of all the infinite elements (A1, A2, A3 ......, B1, B2, B3 ......, C1 & C2, D1 & D2) in Subgroup A, Subgroup B, Subgroup C and Subgroup D above implies the infinity of the list of the consecutive even composites, i.e., 4, 6, 8, 10, 12, 14, 16, 18, 20, 22, etc. to infinity (Subgroup E), which we know to be true in any case, and vice versa, etc. to infinity.

Subgroup A and Subgroup B above are practically "mirror" images of one another - they represent the viewing of the primes and the composites from 2 variant angles - the infinitude, or, finiteness of either implies the infinitude, or, finiteness of the other; the same applies to both Subgroup C and Subgroup D above. It is similar to the following way of viewing a glass which is partially filled: this glass could be described as "half full" or "half empty" if it is half filled, "three-quarter full" or "one-quarter empty" if it is three-quarter filled, or, "one-quarter full" or "three-quarter empty" if it is one-quarter filled, etc.

It is evident that the infinitude, or, finiteness of any one of the above-mentioned elements would imply the infinitude, or, finiteness of the other element that is interlinked with it and vice versa. All these infinite elements are evidently entangled together and complementary, being all the infinite building-blocks of the infinite integers or whole numbers. The infinity of the list of the integers or whole numbers, the primes included, in fact implies that all these various elements within it are infinite, and, vice versa, since all these various elements are closely interlinked and could not do without each other. Therefore, the breaking of the evident intrinsic symmetry of this whole infinite group, i.e., the infinite list of the integers or whole numbers, due to the finiteness of any of the elements within it, could not be possible.

We pose a very important question: Besides questioning whether the infinite list of all the primes pairs separated by 2 integers (twin primes) is really infinite, should we not also be questioning whether the following are really infinite?:

(a) Infinite lists of all the primes pairs separated respectively by 4 integers, 6 integers, 8 integers, 10 integers and sequentially larger integers to infinity (as in Subgroup A above).

(b) Infinite lists of all the respective consecutive odd composites of "infinite sizes" sandwiched between 2 primes (as in Subgroup B above).

(c) The 2 infinite lists with respectively "2 consecutive odd composites separated by 4 integers/1 prime" and "2 consecutive odd composites separated by 6 integers/2 primes" (as in Subgroup C above).

(d) The 2 infinite lists of respective single primes and twin primes separating 2 consecutive odd composites (as in Subgroup D above).

(e) Infinite list of the consecutive even composites all separated by only 2 integers (as in Subgroup E above).

Could there possibly be any symmetry-breaking in the above-mentioned infinite group whence one or more of the elements within it would be finite? In particular, could there be a possibility for the symmetry of this infinite group to be broken due to the finiteness of Element A1 (i.e., the finiteness of the twin primes) within it? Since the above-mentioned group, i.e., the list of the integers or whole numbers, is infinite, it is indeed not possible for all of these elements to be finite. And, there is no evident reason to account for why any of these elements, especially Element A1, i.e., the list of primes separated by 2 integers, or, twin primes, should be finite. In fact, all these infinite elements are like the slabs of various sizes in a building. They are all necessary for the construction of the infinite building known as the "infinite list of the integers or whole numbers" and should thus all be infinite, wherein the symmetry of the infinite group, i.e., the infinite list of the integers or whole numbers, would be preserved.

Therefore, by Lemma 1, all the elements in Subgroup A, Subgroup B, Subgroup C, Subgroup D and Subgroup E above are infinite.

Lemma 2: The Fundamental Theorem of Arithmetic or Unique Factorisation Theorem states that there is only one possible combination of primes which will multiply together to produce any particular composite number, e.g., the only combination of primes which will produce the composite number 2,079 is: 3 x 3 x 3 x 7 x 11. In the same manner, the following composite numbers are also uniquely factorised:

(1) 63 = 3 x 3 x 7 (only)

(2) 153 = 3 x 3 x 17 (only)
(3) 1,021,020 = 2 x 2 x 3 x 5 x 7 x 11 x 13 x 17 (only)

In other words, every positive whole number which is not prime (i.e., every positive whole number which is composite) can be broken up into prime factors, and, this can happen in only 1 way:

$$c \quad = \quad \prod_{p \text{ prime}} p \quad \text{(in only 1 way)}$$

The 10 consecutive twin primes 3 & 5 to 107 & 109, e.g., give rise to the following 10 composite numbers which can be factorised in only 1 way, i.e., can be factorised only by the respective twin primes:

(1) 15 = 3 x 5 (only)
(2) 35 = 5 x 7 (only)
(3) 143 = 11 x 13 (only)
(4) 323 = 17 x 19 (only)
(5) 899 = 29 x 31 (only)
(6) 1,763 = 41 x 43 (only)
(7) 3,599 = 59 x 61 (only)
(8) 5,183 = 71 x 73 (only)
(9) 10,403 = 101 x 103 (only)
(10) 11,663 = 107 x 109 (only)

.

.

.

As the composite numbers are infinite, this implies that there is an infinitude of twin primes acting as prime factors for an infinitude of composite numbers in only 1 way as the twin primes are indispensable, i.e., necessary, as prime factors for the formation of an infinite number of composite numbers which can only be formed through the product of twin primes in only 1 way - the twin primes can never be substituted as prime factors of these composite numbers by other primes.

6 MORE SOLUTIONS FOR THE TWIN PRIMES CONJECTURE

INTRODUCTION

In 1919, Viggo Brun (1885 - 1978) proved that the sum of the reciprocals of the twin primes converges to Brun's constant:

$$1/3 + 1/5 + 1/7 + 1/11 + 1/13 + 1/17 + 1/19 + ... = 1.9021605 ....$$

It is evident that the twin primes thin out as infinity is approached. The problem of whether there is an infinitude of twin primes is an inherently difficult one to solve, as infinity (normally symbolised by: ∞) is a difficult concept and is against common sense. It is impossible to count, calculate or live to infinity, perhaps with the exception of God. Infinity is a nebulous idea and appears to be only an abstraction devoid of any actual practical meaning. How do we quantify infinity? How big is infinity? We could either attempt to prove that the twin primes are finite, or, infinite. If the twin primes were finite, how could we prove that a particular pair of twin primes is the largest existing pair of twin primes, and, if they were infinite, how could we prove that there are always larger and larger pairs of them? It is evidently difficult to prove either, with the former appearing more difficult to prove as the odds seem against it. This chapter provides evidence of the latter, i.e., the infinitude of the twin primes.

PART 1

Theorem:- The twin primes are infinite.

Argument:-

Let 3, 5, 7, 11, 13, 17, 19, …….., n - 2, n be the list of consecutive primes, wherein n & n - 2 are assumed to be the largest existing twin primes pair, within the infinite list of the primes.

Let $3 \times 5 \times 7 \times 11 \times 13 \times 17 \times 19 = a$.

Lemma: (a x …….. x n - 2 x n) - 2, &, (a x …….. x n - 2 x n) - 4 will never be divisible by any of the consecutive primes in the list: 3, 5, 7, 11, 13, 17, 19, …….., n - 2, n, whether they are prime or composite. (See appendix 1.)

This implies that:

If (a x …….. x n - 2 x n) - 2 &/V (a x …….. x n - 2 x n) - 4 are prime, then:

(a x …….. x n - 2 x n) - 2 > (a x …….. x n - 2 x n) - 4 > n > n - 2

If (a x …….. x n - 2 x n) - 2 &/V (a x …….. x n - 2 x n) - 4 are non-prime/composite, then:

(a) each prime factor, e.g., y below, of (a x …….. x n - 2 x n) - 2 > n > n - 2
(b) each prime factor, e.g., z below, of (a x …….. x n - 2 x n) - 4 > n > n - 2

(a x …….. x n - 2 x n) - 2 = prime V composite   (i)

(a x …….. x n - 2 x n) - 4 = prime V composite   (ii)

(i) & (ii) = twin primes, if both (i) & (ii) are prime

(i) & (ii) > n & n - 2

Let Y represent the prime factors of (a x .... x n - 2 x n) - 2 if (a x .... x n - 2 x n) - 2 is not prime (i.e., it is composite), each prime factor may pair up with another prime which differs from it by 2 to form twin primes. Let y = prime factor in Y.

y & y +/- 2 = twin primes, if y +/- 2 is prime

y & y +/- 2 > n & n - 2

Let Z represent the prime factors of (a x .... x n - 2 x n) - 4 if (a x .... x n - 2 x n) - 4 is not prime (i.e., it is composite), each prime factor may pair up with another prime which differs from it by 2 to form twin primes. Let z = prime factor in Z.

z & z +/- 2 = twin primes, if z +/- 2 is prime

z & z +/- 2 > n & n - 2

Therefore: (a x .... x n - 2 x n) - 2 > (a x .... x n - 2 x n) - 4 > y V y +/- 2 V z V z +/- 2 > n > n - 2

By the above, the following, which implies that n & n - 2 are the largest existing twin primes pair, is an impossibility:

n > n - 2 > (a x .... x n - 2 x n) - 2 > (a x .... x n - 2 x n) - 4 > y V y +/- 2 V z V z +/- 2

It is hence clear that no n & n - 2 in any list of consecutive primes can ever possibly be the largest existing twin primes pair, i.e., a largest existing twin primes pair is an impossibility, which implies that the twin primes are infinite. It is possible to find larger twin primes than n & n - 2 no matter how large n & n - 2 are to infinity, with the following formulae involving the list of consecutive primes: (a x .... x n) - 2 & (a x .... x n) - 4 (see appendix 1), which by the nature of their composition are capable of generating new primes/twin primes which will always be larger than n & n - 2. This is an indirect argument or argument by contradiction (reductio ad absurdum) for the infinity of the twin primes (whose mathematical logic is actually the same as that of Euclid's indirect proof of the infinity of the primes), for our assumption of n & n - 2 as the largest existing twin primes pair will be contradicted by the discovery of larger twin primes with these 2 formulae. In fact, by the same principle, all the twin odd integers found between n and (a x .... x n) - 2, which differ from one another by 2 and are not divisible by any of the primes in the list of consecutive primes: 3, 5, 7, 11, 13, 17, 19, ..., n, will be twin primes larger than n & n - 2, our assumed largest existing twin primes pair, which is a contradiction of this assumption, thus proving the infinitude of the twin primes. (Refer to Algorithm 1, as well as Algorithm 2, in Part 5.) In this manner, by continually adding more and more consecutive primes to the list of consecutive primes: 3, 5, 7, 11, 13, 17, 19, ..., n, i.e., continually extending the value of n, and utilising the formula: (a x .... x n) - 2 to perform the computations, many larger and larger twin primes can be found, all the way to infinity, in parallel with the infinitude of the list of consecutive primes: 3, 5, 7, 11, 13, 17, 19, ... of which the twin primes are a part together with other primes pairs, wherein the twin primes are not likely to be finite (as is evident from appendix 2) and can be expected to be infinite. We thus affirm the infinity of the twin primes.

APPENDIX 1

Note: The (only) even prime 2 is omitted from the list of consecutive primes: 3, 5, 7, 11, 13, 17, 19, ..., n - 2, n stated in the chapter, wherein n & n - 2 are assumed to be the largest existing twin primes pair.

The list of newly created primes, and, twin primes for n = 5, 7, 11, 13, 17, 19, ... (n = 19 being the maximum limit achievable with a hand-held calculator) is as follows:-

1] For n = 5, we get the following new primes/new twin primes:

    (3 x 5) - 2 = 13   (X)
    (3 x 5) - 4 = 11   (Y)

2] For n = 7, we get the following new primes/new twin primes:

    (3 x 5 x 7) - 2 =103   (X)
    (3 x 5 x 7) - 4 = 101   (Y)

3] For n = 11, we get the following new primes/new twin primes:

    (3 x 5 x 7 x 11) - 2 = 1,153   (X)
    (3 x 5 x 7 x 11) - 4 = 1,151   (Y)

4] For n = 13, we get the following new prime and composite number with its prime
   factors:

    (3 x 5 x 7 x 11 x 13) - 2 = 15,013   (X) - Prime Number
    (3 x 5 x 7 x 11 x 13) - 4 = 15,011   (Y) - Composite Number (= 17 x 883, with 17 pairing with 19 to form a twin
                                           primes pair and 883 pairing with 881 to form another twin primes pair)

5] For n = 17, we get the following new primes/new twin primes:

    (3 x 5 x 7 x 11 x 13 x 17) - 2 = 255,253   (X)
    (3 x 5 x 7 x 11 x 13 x 17) - 4 = 255,251   (Y)

6] For n = 19, we get the following new prime and composite number with its prime
   factors:

    (3 x 5 x 7 x 11 x 13 x 17 x 19) - 2 = 4,849,843   (X) - Prime Number
    (3 x 5 x 7 x 11 x 13 x 17 x 19) - 4 = 4,849,841   (Y) - Composite Number (= 43 x 112,787, with 43 pairing with 41 to form
                                           a twin primes pair while 112,787 is a stand-alone prime)

            .
            .
            .

Results Of X & Y Above
1) X above generates 6 new primes (13; 103; 1,153; 15,013; 255,253; 4,849,843), nil composite numbers.
2) Y above generates 4 new primes (11; 101; 1,151; 255,251), 2 composite numbers (15,011 = 17 x 883; 4,849,841 = 43 x 112,787).
3) X & Y above together produce 4 pairs of new twin primes (13 & 11; 103 & 101; 1,153 & 1,151; 255,253 & 255,251).
4) The prime factors of X and Y above form 3 pairs of new twin primes with prime partners which differ from them by 2 (19 & 17; 43
   & 41; 883 & 881).
5) All the new twin primes in (3) & (4) above are larger than n & n - 2, the assumed largest existing twin primes pair, which is indirect
   evidence of the infinitude of the twin primes.

Why It Is Impossible For Any n & n - 2 To Be The Largest Existing Twin Primes Pair

$X = (3 \times 5 \times 7 \times 11 \times 13 \times 17 \times 19 \times .... \times n) - 2$, and, $Y = (3 \times 5 \times 7 \times 11 \times 13 \times 17 \times 19 \times .... \times n) - 4$ will never be divisible by any of the consecutive prime numbers in the list: 3, 5, 7, 11, 13, 17, 19, ..., n, whether they are prime or composite (non-prime and divisible by prime numbers or prime factors). This means that none of the consecutive prime numbers in the list: 3, 5, 7, 11, 13, 17, 19, ..., n can ever be factors of X and Y, and, X and Y must be new primes/twin primes larger than all the consecutive prime numbers in the list: 3, 5, 7, 11, 13, 17, 19, ..., n, or, if they were composite (non-prime and divisible by prime numbers or prime factors), their prime factors (and "twin prime" partners which differ from them by 2) must be larger than all the consecutive prime numbers in the list: 3, 5, 7, 11, 13, 17, 19, ..., n. This is a very important mathematical logic, which needs to be grasped in order to understand the argument.

This all implies that no n & n - 2 (if n - 2 were also a prime number) in any list of consecutive prime numbers can ever possibly be the largest existing twin primes pair, since all the new primes/twin primes produced or generated by X and Y will always be larger than n & n - 2. That is, a largest existing twin primes pair is an impossibility, which implies the infinitude of the list of the primes/twin primes.

In other words, by the mathematical logic stated above, which explains why all the new primes/twin primes, which X and Y by the nature of their composition are capable of producing or generating, will always be larger than n & n - 2, no n & n - 2 in any list of consecutive prime numbers: 3, 5, 7, 11, 13, 17, 19, ..., n can ever possibly be the largest existing twin primes pair, i.e., a largest existing twin primes pair is indeed an impossibility, thus implying the infinitude of the list of the twin primes. This is a very important inference.

Regardless of how long the list of the twin primes pairs is, it is possible to find some new twin primes pairs which will always be larger than n & n - 2, our assumed largest existing twin primes pair - the largest twin primes pair in our assumed finite list of the twin primes pairs, with X and Y, which is indirect evidence of the infinity of the twin primes. In fact, by the same principle, all the twin odd integers found between n and X, which differ from one another by 2 and are not divisible by any of the primes in the list of consecutive primes: 3, 5, 7, 11, 13, 17, 19, ..., n, will be twin primes pairs larger than n & n - 2, our assumed largest existing twin primes pair, which is a contradiction of this assumption, hence proving the infinitude of the twin primes. (Refer to Algorithm 1, as well as Algorithm 2, in Part 5.)

APPENDIX 2

Anecdotal Evidence Of The Infinity Of The Twin Primes

<u>TOP TWIN PRIMES IN 2000, 2001, 2007 & 2009</u>

In the year 2000, $4648619711505 \times 2^{60000} \pm 1$ (18,075 digits) had been the top twin primes pair which had been discovered. In the year 2001, it only ranked eighth in the list of top 20 twin primes pairs, with $318032361 \cdot 2^{107001} \pm 1$ (32,220 digits) topping the list. In the year 2007, in the list of top 20 twin primes pairs, $318032361 \cdot 2^{107001} \pm 1$ (32,220 digits) ranked eighth, while $4648619711505 \times 2^{60000} \pm 1$ (18,075 digits) was nowhere to be seen; $2003663613*2^{\wedge}195000-1$ and $2003663613*2^{\wedge}195000+1$ (58,711 digits), which was discovered on January 15, 2007, by Eric Vautier (from France) of the Twin Prime Search (TPS) project in collaboration with PrimeGrid (BOINC platform), was at the top of the list. As at August 2009, $65516468355 \cdot 2^{333333}-1$ and $65516468355 \cdot 2^{333333}+1$ (100,355 digits) is at the top of the list of top 20 twin primes pairs, while $318032361 \cdot 2^{107001} \pm 1$ (32,220 digits) ranks 11th., and, $2003663613*2^{\wedge}195000-1$ and $2003663613*2^{\wedge}195000+1$ (58,711 digits) ranks second in this list.

We can expect larger twin primes than these extremely large twin primes, much larger ones, infinitely larger ones, to be discovered in due course.

LIST OF PRIMES PAIRS FOR THE FIRST 2,500 CONSECUTIVE PRIMES, 2 TO 22,307, RANKED ACCORDING TO THEIR
FREQUENCIES OF APPEARANCE

S. No.	Ranking	Prime Pairs	No. Of Pairs	Percentage
(1)	1	primes pair separated by 6 integers	482	19.29 %
(2)	2	primes pair separated by 4 integers	378	15.13 %

(3)	3	primes pair separated by 2 integers (t. p.)	376	15.05 %
(4)	4	primes pair separated by 12 integers	267	10.68 %
(5)	5	primes pair separated by 10 integers	255	10.20 %
(6)	6	primes pair separated by 8 integers	229	9.16 %
(7)	7	primes pair separated by 14 integers	138	5.52 %
(8)	8	primes pair separated by 18 integers	111	4.44 %
(9)	9	primes pair separated by 16 integers	80	3.20 %
(10)	10	primes pair separated by 20 integers	47	1.88 %
(11)	11	primes pair separated by 22 integers	46	1.84 %
(12)	12	primes pair separated by 30 integers	24	0.96 %
(13)	13	primes pair separated by 28 integers	19	0.76 %
(14)	14	primes pair separated by 24 integers	16	0.64 %
(15)	15	primes pair separated by 26 integers	10	0.40 %
(16)	16	primes pair separated by 34 integers	9	0.36 %
(17)	17	primes pair separated by 36 integers	5	0.20 %
(18)	18	primes pair separated by 32 integers	2	0.08 %
(19)	18	primes pair separated by 40 integers	2	0.08 %
(20)	19	primes pair separated by 42 integers	1	0.04 %
(21)	19	primes pair separated by 52 integers	1	0.04 %

Total No. Of Primes Pairs In List: 2,498

It is evident in the above list that the primes pairs separated by 6 integers, 4 integers and 2 integers (twin primes), among the 21 classifications of primes pairs separated by from 2 integers to 52 integers (primes pairs separated by 38 integers, 44 integers, 46 integers, 48 integers & 50 integers are not among them, but, they are expected to appear further down in the infinite list of the primes), are the most dominant, important. There is a long list, an infinite list, of other primes pairs, besides those shown in the above list, which also play a part as the building-blocks of the infinite list of the integers.

The list of the integers is infinite. The list of the primes is also infinite. The infinite primes are the building-blocks of the infinite integers - the infinite odd integers are all either primes or composites of primes, and, the infinite even integers, except for 2 which is a prime, are all also composites of primes. Therefore, all the primes pairs separated by the integers of various magnitudes, as described above, can never all be finite. If there is any possibility at all for any of these primes pairs to be finite, there is only the possibility that a number of these primes pairs are finite (but never all of them). However, will it have to be the primes pairs separated by 2 integers or twin primes (which are the subject of our investigation here), which are the only primes pair, or, one among a number of primes pairs, which are finite? Why question only the infinity of the primes pairs separated by 2 integers, the twin primes? Are not the infinities of the primes pairs separated by 8 integers and more, whose frequencies of appearance are lower, as compared to those of the primes pairs which are separated by 6, 4 and 2 integers respectively, in the above list of primes pairs, more questionable? Why single out only the twin primes? (There are at least 18 other primes pairs, separated by from 8 integers to 52 integers, whose respective infinities should be more suspect, as is evident from the above list of primes pairs, if any infinities should be doubted. Evidently, the primes pairs separated by 2 integers (twin primes) are not that likely to be finite.)

The above represents anecdotal evidence that the twin primes are infinite, which is a ratification of the actual proof given earlier.

PART 2

Theorem:- There is an infinite number of twin primes.

Numerical Evidence:-
Many believe that the twin primes are infinite. In fact, twin primes pairs could easily be found among the integers. There is evidently no region of the natural number system so remote that it lies beyond the largest twin primes pair. It is even possible to forecast the approximate number of twin primes pairs found in any region of the natural number system.

The occurrence of twin primes pairs is evidently unpredictable or random. This means that the chance of 2 numbers x and x + 2 being prime (twin primes) is somewhat similar to the chance of getting heads on 2 successive tosses of a coin. If 2 successive tosses of a coin are independent, the chance of success of obtaining heads for the 2 successive tosses of the coin is the product of the chances of success of obtaining a head for each toss of the coin. As each coin has probability ½ of coming up heads with a toss, 2 coins would have probability ½ x ½ = ¼ of coming up a pair of heads with a toss.

The prime number theorem, which had been proven, states that if n is a large number, and we select a number x at random between 0 and n, the chance that x is prime would be approximately $1/\log n$, the larger n is, the better would be the approximation given by $1/\log n$ to the proportion of primes in the numbers up to n. Like 2 coins coming up heads, the chance that both x and x + 2 are prime (twin primes) would be approximately $1/(\log n)^2$. That is, there would be approximately $n/(\log n)^2$ twin primes pairs between 0 and n. As n goes to infinity, this fraction approaches infinity. This represents a quantitative version of the twin primes conjecture.

As x + 2 being prime depends on the fact that x is already prime, we should modify the estimate $n/(\log n)^2$ to $(1.32032..)n/(\log n)^2$.

The following is a comparison between the twin primes predicted by the above formula and the twin primes found, where the agreement is evidently very good:-

INTERVAL	TWIN PRIMES	
	PREDICTED	FOUND
100,000,000 - 100,150,000	584	601
1,000,000,000 - 1,000,150,000	461	466
10,000,000,000 - 10,000,150,000	374	389
100,000,000,000 - 100,000,150,000	309	276
1,000,000,000,000 - 1,000,000,150,000	259	276
10,000,000,000,000 - 10,000,000,150,000	221	208
100,000,000,000,000 -	191	186

100,000,000,150,000

1,000,000,000,000,000 -   166   161
1,000,000,000,150,000

All this represents numerical evidence that the twin primes are infinite as we could find more twin primes pairs whenever we look for them.

Argument:-

Lemma: According to the principle of induction in set theory, if a set of natural numbers contains 1, and if it contains n + 1 whenever it contains a number n, then it must contain every natural number, e.g., induction proves that every natural number is a product of primes.

By the above lemma, there is an infinitude of twin primes in the infinite set of the integers; twin primes are found in all the above subsets of integers, from 100,000,000 - 100,150,000 to 1,000,000,000,000,000 - 1,000,000,000,150,000, and many very much larger twin primes have also been found in the infinite set of the integers, e.g., $318032361 \cdot 2^{107001} \pm 1$ (32,220 digits), 2003663613*2^195000-1 and 2003663613*2^195000+1 (58,711 digits), and, $65516468355 \cdot 2^{333333}-1$ and $65516468355 \cdot 2^{333333}+1$ (100,355 digits), to name a few (this long list of twin primes, especially the large twin primes, are obtainable only with the help of modern computer technology) - the principle of induction implies that there must be an infinite number of twin primes, ranging from the twin primes pair, 3 & 5 (smallest twin primes pair), 5 & 7, 11 & 13, 17 & 19, ..., $65516468355 \cdot 2^{333333}-1$ and $65516468355 \cdot 2^{333333}+1$ (100,355 digits - largest twin primes pair discovered as at August 2009), ... upward to infinity, in the infinite set of the integers - in other words, the twin primes conjecture must be true.

PART 3

Theorem:- The list of the twin primes pairs is infinite.

Argument:-

Lemma: According to the precepts of fractal geometry and group theory, symmetry is a very important, intrinsic part of nature. There is symmetry all around us and within us. There is evident symmetry in human bodies, the structures of viruses and bacteria, polymers and ceramic materials, the permutations of numbers, the universe and many others, even the movements of prices in financial markets, the growths of populations, the sound of music, the flow of blood through our circulatory system, the behaviour of people en masse, etc. In other words, regularity, pattern, order, uniformity or symmetry is evident everywhere.

The reasoning here makes use of a very important idea in fractal geometry and group theory, namely, symmetry.

A prime number is an integer which is divisible only by 1 and itself, e.g., 2, 3, 7, 19, etc. A twin primes pair are 2 primes which differ from one another by 2, e.g., 5 & 7, 11 & 13, 17 & 19, and, 29 & 31, etc. A composite number or non-prime is a product of primes or prime factors, e.g., the composite numbers 15 is the product of 2 primes, 3 and 5 (15 = 3 x 5), and 231 is the product of 3 primes, 3, 7 and 11 (231 = 3 x 7 x 11), etc. The integers or whole numbers are either primes or composites and are infinite.

The primes, which Euclid had proven to be infinite, are the atoms or building-blocks of the infinite integers or whole numbers, which comprise of the infinite list of the odd numbers that are all either primes or products of primes (i.e., composites), and, the infinite list of the even numbers that are all products of primes (i.e., composites, with the exception of 2 which is a prime, e.g., 6 = 2 x 3, 8 = 2 x 2 x 2 and 10 = 2 x 5, etc.). The infinite list of the integers or whole numbers may be classified as an infinite group, with various symmetries, subgroups and infinite elements, hidden within it. These various symmetries, subgroups and infinite elements, within this infinite group may be classified as follows:-

(1) <u>Subgroup A</u>: Infinite consecutive primes such as 2, 3, 5, 7, 11, 13, 17, 19, 23, 29, 31, etc. to infinity .... , separated by 2 integers (twin primes), 4 integers, 6 integers, 8 integers, 10 integers, etc. to infinity .... , which, incidentally, except for 2, are all odd numbers; this splitting up of the subgroup into infinite elements is shown below:

   (i) <u>Element A1</u>: Infinite list of all the primes pairs separated by 2 integers (twin primes) .... (Example: 17 & 19)
   (ii) <u>Element A2</u>: Infinite list of all the primes pairs separated by 4 integers/1 odd composite - single composite .... (Example: 79 & 83 separated by 81)
   (iii) <u>Element A3</u>: Infinite list of all the primes pairs separated by 6 integers/2 consecutive odd composites - twin composites .... (Example: 47 & 53 separated by 49 & 51)
   (iv) <u>Element A4</u>: Infinite list of all the primes pairs separated by 8 integers/3 consecutive odd composites - "triple" composites .... (Example: 359 & 367 separated by 361, 363 & 365)
   (v) <u>Element A5</u>: Infinite list of all the primes pairs separated by 10 integers/4 consecutive odd composites - "four-ple" composites .... (Example: 709 & 719 separated by 711, 713, 715 & 717)

.
.
.

(2) <u>Subgroup B</u>: Infinite consecutive odd composites such as 9, 15, 21, 25, 27, 33, 35, 39, 45, 49, etc. to infinity .... , of "infinite sizes" sandwiched between 2 primes; this splitting up of the subgroup into infinite elements is shown below:

   (i) <u>Element B1</u>: Infinite list of all "1 odd composite sandwiched between 2 primes - single composite" .... (Example: 9 sandwiched between the primes 7 & 11)
   (ii) <u>Element B2</u>: Infinite list of all "2 consecutive odd composites sandwiched between 2 primes - twin composites" .... (Example: 253 & 255 sandwiched between the primes 251 & 257)
   (iii) <u>Element B3</u>: Infinite list of all "3 consecutive odd composites sandwiched between 2 primes - "triple" composites" .... (Example: 685, 687 & 689 sandwiched between the primes 683 & 691)
   (iv) <u>Element B4</u>: Infinite list of all "4 consecutive odd composites sandwiched between 2 primes - "four-ple" composites" .... (Example: 2,769, 2,771, 2,773 & 2,775 sandwiched between the primes 2,767 & 2,777)
   (v) <u>Element B5</u>: Infinite list of all "5 consecutive odd composites sandwiched between 2 primes - "five-ple" composites" .... (Example: 19,291, 19,293, 19,295, 19,297 & 19,299 sandwiched between the primes 19,289 & 19,301)

.
.
.

(3) <u>Subgroup C</u>: Infinite consecutive odd composites separated by 4 integers and 6 integers respectively; this splitting up of the subgroup into the 2 infinite elements is shown below:

   (i) <u>Element C1</u>: Infinite list of all "2 consecutive odd composites separated by 4 integers/1 prime" .... (Example: 209 & 213 separated by the prime 211)
   (ii) <u>Element C2</u>: Infinite list of all "2 consecutive odd composites separated by 6 integers/2 primes" .... (Example: 279 & 285 separated by the twin primes 281 & 283)

(4) <u>Subgroup D</u>: Infinite single primes and twin primes separating 2 consecutive odd composites; this splitting up of the subgroup into the 2 infinite elements is shown below:

   (i) <u>Element D1</u>: Infinite list of all the single primes separating 2 consecutive odd composites .... (Example: 23 separating the 2 consecutive odd composites 21 & 25)
   (ii) <u>Element D2</u>: Infinite list of all the twin primes separating 2 consecutive odd composites .... (Example: 11 & 13 separating the 2 consecutive odd composites 9 & 15)

(5) <u>Subgroup E</u>: Infinite consecutive even composites such as 4, 6, 8, 10, 12, 14, 16, 18, 20, 22, etc. to infinity ...., all separated by only 2 integers; this subgroup may be classified as a single infinite element .... There is always 1 even number between a twin primes pair, which is separated by 2 integers, a prime and a composite which are separated by 2 integers, and, 2 composites which are separated by 2 integers. That is, the even numbers are always found in Subgroup A, Subgroup B, Subgroup C and Subgroup D above always evenly spaced out in consecutive order by 2 integers.

There is an evident symmetry in the above-mentioned infinite group, which would be broken if any of the elements within were to be finite. There are close interlinks between all the various infinite elements in all the five subgroups above, e.g., the infinity of the list of all the primes pairs each separated by 6 integers (Element A3/Subgroup A) implies the infinity of the list of all the "2 consecutive odd composites sandwiched between 2 primes - twin composites" (Element B2/Subgroup B) and vice versa, the infinity of the list of all the primes pairs each separated by 2 integers (twin primes) (Element A1/SubgroupA) implies the infinity of the list of all the "2 consecutive odd composites separated by 6 integers/2 primes" (Element C2/Subgroup C) and vice versa, the infinity of the list of all the infinite elements (A1, A2, A3, etc. to infinity) in Subgroup A above, which represents the infinity of the list of the primes which Euclid had in fact proven, implies the infinity of the list of all the "2 consecutive odd composites separated by 4 integers/1 prime" (Element C1/Subgroup C) and vice versa, the infinity of the list of all the "2 consecutive odd composites separated by 6 integers/2 primes" (Element C2/Subgroup C) implies the infinity of the list of all the twin primes separating 2 consecutive odd composites (Element D2/Subgroup D) and vice versa, the infinity of all the lists of all the infinite elements (A1, A2, A3 ......, B1, B2, B3 ......, C1 & C2, D1 & D2) in Subgroup A, Subgroup B, Subgroup C and Subgroup D above implies the infinity of the list of the consecutive even composites, i.e., 4, 6, 8, 10, 12, 14, 16, 18, 20, 22, etc. to infinity (Subgroup E), which we know to be true in any case, and vice versa, etc. to infinity.

Subgroup A and Subgroup B above are practically "mirror" images of one another - they represent the viewing of the primes and the composites from 2 variant angles - the infinitude, or, finiteness of either implies the infinitude, or, finiteness of the other; the same applies to both Subgroup C and Subgroup D above. It is similar to the following way of viewing a glass which is partially filled: this glass could be described as "half full" or "half empty" if it is half filled, "three-quarter full" or "one-quarter empty" if it is three-quarter filled, or, "one-quarter full" or "three-quarter empty" if it is one-quarter filled, etc.

It is evident that the infinitude, or, finiteness of any one of the above-mentioned elements would imply the infinitude, or, finiteness of the other element that is interlinked with it and vice versa. All these infinite elements are evidently entangled together and complementary, being all the infinite building-blocks of the infinite integers or whole numbers. The infinity of the list of the integers or whole numbers, the primes included, in fact implies that all these various elements within it are infinite, and, vice versa, since all these various elements are closely interlinked and could not do without each other. Therefore, the breaking of the evident intrinsic symmetry of this whole infinite group, i.e., the infinite list of the integers or whole numbers, due to the finiteness of any of the elements within it, could not be possible.

We pose a very important question: Besides questioning whether the infinite list of all the primes pairs separated by 2 integers (twin primes) is really infinite, should we not also be questioning whether the following are really infinite?:

(a) Infinite lists of all the primes pairs separated respectively by 4 integers, 6 integers, 8 integers, 10 integers and sequentially larger integers to infinity (as in Subgroup A above).

(b) Infinite lists of all the respective consecutive odd composites of "infinite sizes" sandwiched between 2 primes (as in Subgroup B above).

(c) The 2 infinite lists with respectively "2 consecutive odd composites separated by 4 integers/1 prime" and "2 consecutive odd composites separated by 6 integers/2 primes" (as in Subgroup C above).

(d) The 2 infinite lists of respective single primes and twin primes separating 2 consecutive odd composites (as in Subgroup D above).

(e) Infinite list of the consecutive even composites all separated by only 2 integers (as in Subgroup E above).

Could there possibly be any symmetry-breaking in the above-mentioned infinite group whence one or more of the elements within it would be finite? In particular, could there be a possibility for the symmetry of this infinite group to be broken due to the finiteness of Element A1 (i.e., the finiteness of the twin primes) within it? Since the above-mentioned group, i.e., the list of the integers or whole numbers, is infinite, it is indeed not possible for all of these elements to be finite. And, there is no evident reason to account for why any of these elements, especially Element A1, i.e., the list of primes separated by 2 integers, or, twin primes, should be finite. In fact, all these infinite elements are like the slabs of various sizes in a building. They are all necessary for the construction of the infinite building known as the "infinite list of the integers or whole numbers" and should thus all be infinite, wherein the symmetry of the infinite group, i.e., the infinite list of the integers or whole numbers, would be preserved. The following explanation should make it clear that all these elements in the above-mentioned infinite group have to be infinite.

We here examine a hypothetical case wherein one of the elements within the infinite group is finite. Let us assume that Element A1, the list of primes separated by 2 integers, or, twin primes, in Subgroup A above, which is our target object, is finite. Let us look at the following list of consecutive integers or whole numbers where the bolded numbers are primes and the bolded, italised numbers are twin primes:-

.... 201, 202, 203, 204, 205, 206, 207, 208, 209, 210, **211**, 212, 213, 214, 215, 216, 217, 218, 219,

220, 221, 222, **223**, 224, 225, 226, **227**, 228, **229**, 230, 231, 232, **233**, 234, 235, 236, 237, 238,

239, 240, **241**, 242, 243, 244, 245, 246, 247, 248, 249, 250, **251**, 252, 253, 254, 255, 256, **257**,

258, 259, 260, 261, 262, **263**, 264, 265, 266, 267, 268, **269**, 270, **271**, 272, 273, 274, 275, 276,

277, 278, 279, 280, **281**, 282, **283**, 284, 285, 286, 287, 288, 289, 290, 291, 292, **293**, 294, 295, 296, 297,

298, 299, 300, 301 ....

Now, what would happen if the list of the twin primes (Element A1) were finite? Let us, e.g., assume that 269 & 271 are the largest existing twin primes pair - there are no twin primes larger than they. And let us look at the above list of consecutive integers or whole numbers now:-

.... 201, 202, 203, 204, 205, 206, 207, 208, 209, 210, **211**, 212, 213, 214, 215, 216, 217, 218, 219,

220, 221, 222, **223**, 224, 225, 226, **227**, 228, **229**, 230, 231, 232, **233**, 234, 235, 236, 237, 238,

239, 240, **241**, 242, 243, 244, 245, 246, 247, 248, 249, 250, **251**, 252, 253, 254, 255, 256, **257**,

258, 259, 260, 261, 262, **263**, 264, 265, 266, 267, 268, **269**, 270, **271**, 272, 273, 274, 275, 276,

277, 278, 279, 280,   , 282,   , 284, 285, 286, 287, 288, 289, 290, 291, 292, **293**, 294, 295, 296, 297,

298, 299, 300, 301 ....

Since 269 & 271 are the largest existing twin primes, the twin primes pair, 281 & 283, would not exist anymore, resulting in 2 gaps (between the integers 280 and 284) in the above list of consecutive integers or whole numbers. Thus, the earlier complete list of consecutive integers or whole numbers has become incomplete and discontinuous as a result - 2 building-blocks, 281 and 283, which are twin primes (primes separated by 2 integers), are missing from this list now - the list has now become asymmetrical, i.e., symmetry-breaking has occurred. As the list of integers or whole numbers is always complete, continuous, infinite and symmetrical, such a list with 2 missing consecutive odd numbers which are both primes (these missing consecutive odd numbers are twin primes) is an absurdity - it could never exist. Therefore, a largest existing twin primes pair, such as 269 & 271 mentioned above, is an absurdity - a largest existing twin primes pair could never exist, and, it is evident here that, of necessity, in order that the list of the integers or whole numbers would be complete, continuous, infinite and symmetrical, the list of the twin primes (primes separated by 2 integers), as well as the lists of the primes pairs separated by 4 integers, 6 integers, 8 integers, 10 integers and so on in arithmetically ascending order to infinity, have to be infinite. That is, all the elements in Subgroup A above have to be infinite.

By a similar reasoning, the same has to also apply to all the elements in Subgroup B, Subgroup C, Subgroup D and Subgroup E above.

Therefore, by the above lemma, all the elements in Subgroup A, Subgroup B, Subgroup C, Subgroup D and Subgroup E above are infinite.

It is clear that the twin primes are infinite.

PART 4

Theorem:- The twin primes are infinite.

Argument:-

Lemma: A fraction of infinity is also infinite. (See appendix.)

The list of the primes, which are the building-blocks of the integers, had been proven by Euclid to be infinite. The question now is whether the list of the twin primes is also infinite.

It would be appropriate to conduct an examination of the primes pairs separated by 2 integers (commonly known as twin primes) vis-a-vis the other primes pairs separated by more than 2 integers, e.g., 4 integers, 6 integers, 8 integers and more. For this purpose, we select a reasonably large list of consecutive primes, which may be regarded as a basic unit of the infinite list of the primes; we examine a compilation of such data obtained from, say, the list of the first 2,500 consecutive primes - 2 to 22,307 - which is as follows:-

LIST OF PRIMES PAIRS FOR THE FIRST 2,500 CONSECUTIVE PRIMES, 2 TO 22,307, RANKED ACCORDING TO THEIR FREQUENCIES OF APPEARANCE

S. No.	Ranking	Prime Pairs	No. Of Pairs	Percentage
(1)	1	primes pair separated by 6 integers	482	19.29 %
(2)	2	primes pair separated by 4 integers	378	15.13 %
(3)	3	primes pair separated by 2 integers (t. p.)	376	15.05 %
(4)	4	primes pair separated by 12 integers	267	10.68 %
(5)	5	primes pair separated by 10 integers	255	10.20 %
(6)	6	primes pair separated by 8 integers	229	9.16 %
(7)	7	primes pair separated by 14 integers	138	5.52 %
(8)	8	primes pair separated by 18 integers	111	4.44 %
(9)	9	primes pair separated by 16 integers	80	3.20 %
(10)	10	primes pair separated by 20 integers	47	1.88 %
(11)	11	primes pair separated by 22 integers	46	1.84 %
(12)	12	primes pair separated by 30 integers	24	0.96 %
(13)	13	primes pair separated by 28 integers	19	0.76 %
(14)	14	primes pair separated by 24 integers	16	0.64 %
(15)	15	primes pair separated by 26 integers	10	0.40 %
(16)	16	primes pair separated by 34 integers	9	0.36 %
(17)	17	primes pair separated by 36 integers	5	0.20 %
(18)	18	primes pair separated by 32 integers	2	0.08 %
(19)	18	primes pair separated by 40 integers	2	0.08 %
(20)	19	primes pair separated by 42 integers	1	0.04 %
(21)	19	primes pair separated by 52 integers	1	0.04 %

Total No. Of Primes Pairs In List: 2,498

It is evident in the above list that the primes pairs separated by 6 integers, 4 integers and 2 integers (twin primes), among the 21 classifications of primes pairs separated by from 2 integers to 52 integers (primes pairs separated by 38 integers, 44 integers, 46 integers, 48 integers & 50 integers are not among them, but, they are expected to appear further down in the infinite list of the primes), are the most dominant, important. There is a long list, an infinite list, of other primes pairs, besides those shown in the above list, which also play a part as the building-blocks of the infinite list of the integers. We shall prove the infinity of all these various building-blocks below.

The list of the integers is infinite. The list of the primes is also infinite. The infinite primes are the building-blocks of the infinite integers - the infinite odd integers are all either primes or composites of primes, and, the infinite even integers, except for 2 which is a prime, are all also composites of primes. Therefore, all the primes pairs separated by the integers of various magnitudes, as described above, could never all be finite. If there is any possibility at all for any of these primes pairs to be finite, there is only the possibility that a number of these primes pairs are finite (but never all of them). However, would it have to be the primes pairs separated by 2 integers or twin primes (which are the subject of our investigation here), which are the only primes pair, or, one among a number of primes pairs, which are finite? Why question only the infinity of the primes pairs separated by 2 integers, the twin primes? Are not the infinities of the primes pairs separated by 8 integers and more, whose frequencies of appearance are lower, as compared to those of the primes pairs which are separated by 6, 4 and 2 integers respectively, in the above list of primes pairs, more questionable? Why single out only the twin primes? (There are at least 18 other primes pairs, separated by from 8 integers to 52 integers, whose respective infinities should be more suspect, as is evident from the above list of primes pairs, if any infinities should be doubted. It is evident that the primes pairs separated by 2 integers (twin primes) are not that likely to be finite.)

All of the above lists of primes pairs which are respectively separated by from 2 integers to 52 integers are each respectively a fraction of the infinite list of the primes: the list of primes pairs separated by 2 integers (twin primes) is a fraction of the infinite list of the primes, the list of primes pairs separated by 4 integers is also a fraction of the infinite list of the primes, the list of primes pairs separated by 6 integers is a fraction of the infinite list of the primes too, and so on all the way down to the list of primes pairs separated by 52 integers, and beyond to infinity.

Therefore, by the above lemma, all these various lists of primes pairs separated by integers of various magnitudes are each also infinite.

APPENDIX - EUCLID'S PROOF OF THE INFINITY OF THE PYTHAGOREAN TRIPLES: A FRACTION OF INFINITY IS ALSO INFINITE

A Pythagorean triple is a set of 3 integers wherein 1 number squared added to another number squared equals the third number squared, e.g., $3^2 (9) + 4^2 (16) = 5^2 (25)$ below.

Euclid's proof of the infinity of the Pythagorean triples begins with the statement that the difference between 2 successive square numbers is always an odd number, as is evident below:

Column 1	Column 2	Column 3
1^2	1	
		} 3 (difference between 4 & 1)
2^2	4	
		} 5 (difference between 9 & 4)
3^2	9	
		} 7 (difference between 16 & 9)
4^2	16	
		} 9 (difference between 25 & 16)
5^2	25	
		} 11 (difference between 36 & 25)
6^2	36	
		} 13 (difference between 49 & 36)
7^2	49	
		} 15 (difference between 64 & 49)
8^2	64	
		} 17 (difference between 81 & 64)
9^2	81	
		} 19 (difference between 100 & 81)
10^2	100	
.	.	.
.	.	.
.	.	.

Every one of the infinity of odd numbers (in Column 3 above) could be added to a particular square number (in Column 2 above) to give another square number, e.g., 3 in Column 3 above could be added to 1 in Column 2 above to give the square number 4, 5 in Column 3 above could be added to 4 in Column 2 above to give the square number 9, 7 in Column 3 above could be added to 9 in Column 2 above to give the square number 16, and so on .... A fraction of the infinite odd numbers in Column 3 above are themselves square numbers, e.g., the odd number 9 in Column 3 above is a square number, and, is the only square number in the list of odd numbers shown there, representing there a fraction of the infinite odd numbers which are square numbers. However, a fraction of infinity is also infinite.

Therefore, there are also an infinity of odd square numbers (in Column 3 above) which could each be added to another square number (in Column 2 above) to give another square number, i.e., there is an infinitude of Pythagorean triples.

PART 5

Theorem:- The twin primes are infinite.

Numerical Evidence:-
The following algorithms would be able to generate or sieve all the twin primes in any range of odd numbers which are all larger than those in the list of known consecutive primes/twin primes; these 2 important algorithms would provide plenty of numerical evidence that the twin primes are infinite:-

Algorithm 1
We would provide an example with Items (1) to (3) from the following list of products of consecutive primes/twin primes, which should be sufficient for our purpose here:-

1) $3 \times 5 = 15$
2) $3 \times 5 \times 7 = 105$
3) $3 \times 5 \times 7 \times 11 = 1,155$
4) $3 \times 5 \times 7 \times 11 \times 13 = 15,015$
5) $3 \times 5 \times 7 \times 11 \times 13 \times 17 = 255,255$
6) $3 \times 5 \times 7 \times 11 \times 13 \times 17 \times 19 = 4,849,845$

.
.
.

The example is as follows:-

1) For $3 \times 5 = 15$, we would find all the consecutive pairs of odd numbers between 5 & 15 which differ from one another by 2 and are not divisible by any of the consecutive primes/twin primes 3 & 5 in the list of consecutive primes/twin primes 3×5 whose product is 15.

   There is only 1 pair of odd numbers between 5 & 15 which differ from one another by 2 and are not divisible by the consecutive primes/twin primes 3 & 5 in the list of consecutive primes/twin primes 3×5 - they are the twin primes 11 & 13.

2) Similarly, for $3 \times 5 \times 7 = 105$, we would find all the consecutive pairs of odd numbers between 7 & 105 which differ from one another by 2 and are not divisible by any of the consecutive primes/twin primes 3, 5 & 7 in the list of consecutive primes/twin primes $3 \times 5 \times 7$ whose product is 105.

   The consecutive pairs of odd numbers between 7 & 105 which differ from one another by 2 and are not divisible by the consecutive primes/twin primes 3, 5 & 7 are the following consecutive twin primes:

   (a) 11 & 13
   (b) 17 & 19
   (c) 29 & 31
   (d) 41 & 43
   (e) 59 & 61
   (f) 71 & 73
   (g) 101 & 103

3) Similarly, in this final case, for 3 x 5 x 7 x 11 = 1,155, we would find all the consecutive pairs of odd numbers between 11 & 1,155 which differ from one another by 2 and are not divisible by any of the consecutive primes/twin primes 3, 5, 7 & 11 in the list of consecutive primes/twin primes 3 x 5 x 7 x 11 whose product is 1,155.

The consecutive pairs of odd numbers between 11 & 1,155 which differ from one another by 2 and are not divisible by the consecutive primes/twin primes 3, 5, 7 & 11 are consecutive twin primes, some of which are as follows:

(a) 17 & 19
(b) 29 & 31
(c) 41 & 43
(d) 59 & 61
(e) 71 & 73
(f) 101 & 103
(g) 107 & 109
(h) 137 & 139
(i) 149 & 151
(j) 179 & 181
(k) Etc. to 1,151 & 1,153

In this way, we would also be able to achieve the following:-

1) For 3 x 5 x 7 x 11 x 13 = 15,015, find all the consecutive twin primes between 13 and 15,015.
2) For 3 x 5 x 7 x 11 x 13 x 17 = 255,255, find all the consecutive twin primes between 17 and 255,255.
3) For 3 x 5 x 7 x 11 x 13 x 17 x 19 = 4,849,845, find all the consecutive twin primes between 19 and 4,849,845.

Algorithm 2

We would, similar to Algorithm 1 above, also provide an example with Items (1) to (3) from the following list of products of consecutive primes/twin primes, which should be sufficient for our purpose here:-

1) 3 x 5 = 15
2) 3 x 5 x 7 = 105
3) 3 x 5 x 7 x 11 = 1,155
4) 3 x 5 x 7 x 11 x 13 = 15, 015
5) 3 x 5 x 7 x 11 x 13 x 17 = 255,255
6) 3 x 5 x 7 x 11 x 13 x 17 x 19 = 4,849,845

The example is as follows:-

1) For 3 x 5 = 15, we would first find all the consecutive pairs of even numbers between 5 & 15 which differ from one another by 2 and are not divisible by any of the consecutive primes/twin primes 3 & 5 in the list of consecutive primes/twin primes 3 x 5. Then we deduct each of these consecutive pairs of even numbers which are not divisible by any of the consecutive primes/twin primes 3 & 5 from the product of these consecutive primes/twin primes 3 x 5 which is 15. The results would each be 1 pair of twin primes, 1 prime & 1 composite of primes, or, 2 composites of primes. In this way, we would be able to find all the consecutive twin primes between 5 & 15.

There is only 1 pair of even numbers between 5 & 15 which differ from one another by 2 and are not divisible by any of the consecutive primes/twin primes 3 & 5 in the list of consecutive primes/twin primes 3 x 5 - they are the pair 2 & 4.

The following is the result after we deduct this pair of even numbers 2 & 4 which are not divisible by any of the consecutive primes/twin primes 3 & 5 from the product of these consecutive primes/twin primes 3 x 5 which is 15:

(a) 15 - 2 & 15 - 4: 13 & 11 (twin primes)

2) Similarly, for 3 x 5 x 7 = 105, we would first find all the consecutive pairs of even numbers between 7 & 105 which differ from one another by 2 and are not divisible by any of the consecutive primes/twin primes 3, 5 & 7 in the list of consecutive primes/twin primes 3 x 5 x 7, which are as follows:

(a) 2 & 4
(b) 32 & 34
(c) 44 & 46
(d) 62 & 64
(e) 74 & 76
(f) 86 & 88
(g) 92 & 94

Then we deduct each of these consecutive pairs of even numbers which are not divisible by any of the consecutive primes/twin primes 3, 5 & 7 from the product of these consecutive primes/twin primes 3 x 5 x 7 which is 105. The results would each be 1 pair of twin primes, 1 prime & 1 composite of primes, or, 2 composites of primes. In this way, we would be able to find all the consecutive twin primes between 7 & 105, which are as follows:

(a) 105 - 2 & 105 - 4:   103 & 101 (twin primes)
(b) 105 - 32 & 105 - 34: 73 & 71 (twin primes)
(c) 105 - 44 & 105 - 46: 61 & 59 (twin primes)
(d) 105 - 62 & 105 - 64: 43 & 41 (twin primes)
(e) 105 - 74 & 105 - 76: 31 & 29 (twin primes)
(f) 105 - 86 & 105 - 88: 19 & 17 (twin primes)
(g) 105 - 92 & 105 - 94: 13 & 11 (twin primes)

3) Similarly, in this final case, for 3 x 5 x 7 x 11 = 1,155, we would first find all the consecutive pairs of even numbers between 11 & 1,155 which differ from one another by 2 and are not divisible by any of the consecutive primes/twin primes 3, 5, 7 & 11 in the list of consecutive primes/twin primes 3 x 5 x 7 x 11, some of which are as follows:

(a) 2 & 4
(b) 32 & 34
(c) 62 & 64
(d) 74 & 76
(e) 92 & 94
(f) 116 & 118
(g) 122 & 124
(h) 134 & 136
(i) Etc. to 1,136 & 1,138

Next we deduct each of these consecutive pairs of even numbers which are not divisible by any of the consecutive primes/twin primes 3, 5, 7 & 11 from the product of these consecutive primes/twin primes 3 x 5 x 7 x 11 which is 1,155. The results would each be 1 pair of twin primes, 1 prime & 1 composite of primes, or, 2 composites of primes. In this way, we would be able to find all the consecutive twin primes between 11 & 1,155, some of which are as follows:

(a) 1,155 - 2 & 1,155 - 4:   1,153 & 1,151 (twin primes)

(b) 1,155 - 32 & 1,155 - 34:   1,123 (prime) & 1,121 (composite of primes which are each larger than 3, 5, 7 & 11 = 19 x 59)

(c) 1,155 - 62 & 1,155 - 64:   1,093 & 1,091 (twin primes)

(d) 1,155 - 74 & 1,155 - 76:   1,081 & 1,079 (composite of primes which are each larger than 3, 5, 7 & 11 = 23 x 47) (composite of primes which are each larger than 3, 5, 7 & 11 = 13 x 83)

(e) 1,155 - 92 & 1,155 - 94:   1,063 & 1,061 (twin primes)

(f) 1,155 - 116 & 1,155 - 118:   1,039 (prime) & 1,037 (composite of primes which are each larger than 3, 5, 7 & 11 = 17 x 61)

(g) 1,155 - 122 & 1,155 - 124:   1,033 & 1,031 (twin primes)

(h) 1,155 - 134 & 1,155 - 136:   1,021 & 1,019 (twin primes)

(i) Etc. to 1,155 - 1,136 & 1,155 - 1,138:   19 & 17 (twin primes)

In like manner, we would also be able to achieve the following:-

1) For 3 x 5 x 7 x 11 x 13 = 15,015, find all the consecutive twin primes between 13 and 15,015.
2) For 3 x 5 x 7 x 11 x 13 x 17 = 255,255, find all the consecutive twin primes between 17 and 255,255.
3) For 3 x 5 x 7 x 11 x 13 x 17 x 19 = 4,849,845, find all the consecutive twin primes between 19 and 4,849,845.

.

.

By utilising any of the above algorithms, we would be able to find many twin primes which are all larger than those in any chosen list of consecutive primes/twin primes, i.e., we would be able to generate many larger and larger twin primes with these algorithms.

Argument:-

Lemma: According to the principle of induction in set theory, if a set of natural numbers contains 1, and if it contains n + 1 whenever it contains a number n, then it must contain every natural number, e.g., induction proves that every natural number is a product of primes.

We could have a "feel" of the infinity of the twin primes when we look at the lists of the top twin primes discovered to-date, which constitutes an argument for the twin primes' infinity; incidentally, this was more or less the kind of proof that had been used to settle the Four-Colour problem in 1976 by Wolfgang Haken and Kenneth Appel, which relied on the computing powers of powerful computers - 1,200 hours of computer time, in fact. Both Haken and Appel had studied the work of Heinrich Heesch who had claimed that the infinity of infinitely variable maps could be constructed from some finite number of finite maps and that by studying these "building-block" maps (these basic maps being the equivalent of the electron, proton and neutron, the fundamental objects from which all else could be constructed) it could be possible to get a hold on the general problem. Using the same reasoning Haken and Appel then reduced the Four-Colour problem to 1,482 "building-block" configurations, whereby proving that these 1,482 maps were four-colourable would imply that all maps would be four-colourable. Likewise, the twin primes' infinitude could be regarded as proven if a pair or more of twin primes were discovered which exceed a certain designated finite limit, a limit which is so expansive that it is indescribable, unimaginable and unnamable, such that it could be regarded as representing infinity itself - the pair of twin primes which represents this limit would be in effect a pair of googolplexes and virtually the point or focus at infinity. (Here, the designated finite limit could be treated as approximately equal to infinity.) According to the approximation theorem (20th. century), any two numbers are approximately equal - this holds whenever the range of approximation is greater than the modulus (positive value) of the difference of the numbers. In geometry, infinity is regarded as a "location", a finite entity, for instance, parallel lines could be said to intersect at a point at infinity, and, parallel planes at a line at infinity; the asymptote to a curve could be regarded as intersecting the curve at infinity. The idea of infinity as a location had been introduced by Johann Kepler, who pointed out that a parabola could be regarded as an ellipse or a hyperbola with one focus at infinity. The idea had been developed by Girard Desargues in his formulation of projective geometry, which assumed the existence of an ideal point at infinity. This equivalency link between infinity and finiteness also appears in infinite series, which are series with an unlimited number of terms. Infinite series could be either divergent or convergent. Convergence is an important feature of a series, i.e., series that are convergent play a major role in mathematics. (It is thus necessary to be able to test a series for convergence by such tests as the Cauchy convergence test, the Cauchy integral test, Abel's test and Dirichlet's test.) An infinite series which converges to a finite sum is known as a convergent series.

We could thus prove the infinitude of the twin primes by using any of the above algorithms, preferably the evidently more efficient Algorithm 1, with the help of powerful computers, to find a pair or more of googolplex twin primes beyond the designated point or focus at infinity which is also a pair of googolplex twin primes, as is described above. (As at August 2009, $65516468355 \cdot 2^{333333}-1$ and $65516468355 \cdot 2^{333333}+1$ (100,355 digits) is at the top of the list of top 20 twin primes pairs, while $2003663613*2^{195000}-1$ and $2003663613*2^{195000}+1$ (58,711 digits), which was the top twin primes pair in 2007, ranks second in this list. By utilising any of the above algorithms, preferably Algorithm 1, and the aid of powerful computers it would be possible to find larger and larger twin primes than these.) This would be a computer-assisted proof similar to the above-described proof of the Four-Colour problem in 1976 by Wolfgang Haken and Kenneth Appel. By the above lemma, the discovery of these larger googolplex twin primes beyond the above-said point or focus at infinity would imply that the twin primes are infinite. This would be a constructive, substantiated proof of the infinity of the twin primes. It would evidently be difficult to accept a proof of the twin primes conjecture without having to confirm or check the validity of the logic by computing a sufficiently long list of twin primes, even to the extent of looking out for counter-examples. Hence, the great importance of the above algorithms.

Also, we could have an indirect proof, a proof by contradiction, for no matter how large the twin primes we discover with any of the 2 algorithms are, we could find more twin primes with any of the 2 algorithms which would always be larger than the last pair of twin primes we discovered - all these newly found twin primes would be evidently larger than all the primes and twin primes in the list of products of consecutive primes/twin primes utilised to generate these new twin primes. (Refer to Part 1.)

All this would constitute a well substantiated constructive proof of the infinity of the twin primes.

CONCLUSION

An array of methods has been adopted in this chapter in solving the twin primes problem.

The inductive method, which is a well-established proof, is one of the methods utilised. The following lends support to this inductive argument for the infinity of the twin primes: (a) The characteristic of a mountain or infinite volume of sand is reflected in the characteristic of some grains of sand found there so that studying the characteristic of some grains of sand found there is enough for deducing the characteristic of the mountain or infinite volume of sand, to ascertain the quality of a batch of products it is only necessary to inspect some carefully selected samples from that batch of products and not everyone of the products and to carry out a population census, i.e., find out the characteristics of a population, it is only necessary to carry out a survey on some carefully selected respondents and not the whole population; in like manner, by the same principle, we just need to study a carefully selected list of integers and their associated primes and twin primes and deduce by induction whether the twin primes would always turn up, appear infinitely, in the list which is itself infinite - this act is rather like extrapolation. (For example, there are 376 pairs of twin primes (752 primes) found within the 2,500 consecutive primes from 2 to 22,307 - this means that 30.08%, which is sizeable, of the 2,500, not a small quantity, consecutive primes are twin primes. 3, 5 & 7 are the only "triple" primes found. There is no regularity in pattern in the appearance of the twin primes, except that the intervals between consecutive twin primes vary greatly by from 4 integers to 370 integers - the intervals between the consecutive twin primes increase and decrease, and, then increase and decrease again, by turns, giving rise to a graph that is characterised by many peaks, i.e., the curve is rough and nonlinear, making its description (hence, forecast of the twin primes) by differential equations practically impossible. By the principle of induction in this case we could deduce that the twin primes would be infinite.) (b) Therefore, if x is a subset of y and if x is a list of prime numbers while y is another list of prime numbers, the characteristic presence of twin primes in x suggests the characteristic presence of twin primes in y, so that if y is an infinite list of prime numbers, whence the prime numbers in it run to infinity, so do the twin primes in it. In fact, induction plays an important part in a number of the arguments.

The other argument used to prove the twin primes' infinity is the indirect (reductio ad absurdum) method, which had been used by Euclid and other mathematicians after him. Logically, 1 or 2 examples of "contradiction" should be sufficient proof of infinity, for it does not make sense to have a need for an infinite number of cases of "contradiction", as our proof would then have to be infinitely and impossibly long, an absurdity. This method of proof is "proof by implication" as a result of "contradiction" - which is a "short-cut" and smart way in proving infinity, instead of "proving infinity by counting to infinity", which is ludicrous, and, impossible. Hence, 1 or 2 cases of "contradiction" should be sufficient for implying that there would be an infinitude of twin primes, which of course also tacitly implies that there would be an infinitude of the number of cases of such "contradiction". (Euclid evidently had this logical point in mind when he formulated the indirect (reductio ad absurdum) proof of the infinity of the primes.) This method of proof had been cleverly used by a number of mathematicians, not the least by the great German mathematician, David Hilbert. For example, Hilbert had used an indirect method (the "reductio ad absurdum" proof) to prove Gordan's Theorem without having to show an actual "construction", a proof which had been accepted by his peers.

There is also the involvement of concepts from set theory, group theory, geometry, etc.

The chapter describes several ways of finding or generating twin primes. Importantly, it presents 2 algorithms for generating or sieving all the twin primes in any range of odd numbers - by utilising any of these 2 algorithms, we would be able to find many twin primes which are all larger than those in any chosen list of consecutive primes, i.e., we would be able to generate many larger and larger twin primes. This is indeed significant. There is evidently some deep meaning in the ease with which the twin primes turn up, as is manifest in the above-described ways of obtaining them. It is thus evident that the twin primes are an inherent characteristic of the infinite prime numbers (as well as odd numbers), a characteristic which could be regarded as "self-similar" or "fractal". A twin primes pair is in effect any pair of odd numbers which differ from one another by 2 and are indivisible by any number except itself, the negative of itself, +1 and -1 (i.e., the pair of odd numbers are prime numbers). Any consecutive odd numbers or odd numbers that differ from one another by 2 are therefore potential prime numbers, as well as potential twin primes, and, the likelihood of them being prime is infinite (vide Euclid's proof and Dirichlet's Theorem), i.e., the primes would always be found amongst them and would be there all the way to infinity (the primes being evidently the "atoms" or building-blocks of all the whole numbers or integers, i.e., all the odd numbers and even numbers - every odd number or integer is either a prime number or composite of prime numbers (i.e., the integer has prime factors), and, every even number is the sum of two prime numbers (vide the Goldbach conjecture which, it appears, practically all mathematicians believe to be true), as well as the product of prime numbers (composite)); hence, the likelihood of them being twin primes is infinite as well (the twin primes being an inherent property of the infinite prime numbers - as well as odd numbers).

So far, there has not been any indication or confirmation that the number of twin primes is finite and the so-called largest existing pair of twin primes has not been found and confirmed (which of course would be impossible to find and confirm if the twin primes were infinite). On the other hand, practically everyone could intuit that the number of twin primes is infinite. Furthermore, there are much evidences or indications that there is an infinitude of twin primes.

We thus conclude that the list of the twin primes is infinite.

7 SOME SOLUTIONS FOR THE GOLDBACH CONJECTURE

The expected mode of solving the Goldbach conjecture appears to be the utilisation of advanced calculus or analysis, e.g., by the summation, or, integration, of the reciprocals involving directly or indirectly the primes to see whether they converge or diverge, in order to get a "feel" of the pattern of the distribution of the primes. But, such a method of solving the problem has evidently not succeeded so far. Some other approach or approaches could be more appropriate. A number of such approaches are brought up here.

INTRODUCTION

The problem of whether there is an infinitude of cases of even numbers which are each the sum of 2 primes is an inherently difficult one to solve, as infinity (normally symbolised by: ∞) is a difficult concept and is against common sense. It is impossible to count, calculate or live to infinity, perhaps with the exception of God. Infinity is a nebulous idea and appears to be only an abstraction devoid of any actual practical meaning. How do we quantify infinity? How big is infinity? The difficulty of the problem of infinity has been compounded by Georg Cantor who proved that there are actually different sizes to infinity, an idea so bizarre to many mathematicians that he was attacked for his ideas during much of his career. The attack was so serious that he suffered mental illness and severe depression. However, after his death his ideas became widely accepted as the only consistent, accurate and powerful definition of infinity. Hilbert had honoured him by saying, "No one shall drive us from the paradise Cantor has created for us." Nevertheless, in this chapter offering solutions for infinity, in this case the infinity of the even numbers which are each the sum of 2 primes, incontrovertible evidence that the peculiar characteristics of the prime numbers themselves among other things contribute to the infinite "generation" of such even numbers would be put forward.

PART 1

Theorem:- Every even number after 2 is the sum of 2 primes.
Solution:-
Christian Goldbach, tutor to the teenage Czar Peter II, had examined dozens of even numbers and noticed that he could split all of them into the sum of 2 primes. Thus, his conjecture that every even number after the number 2 is the sum of 2 prime numbers, for example:

$$4 = 2 + 2$$
$$6 = 3 + 3$$
$$8 = 3 + 5$$
$$10 = 3 + 7 \text{ and } 5 + 5$$
$$12 = 5 + 7$$
$$14 = 3 + 11 \text{ and } 7 + 7$$
$$16 = 3 + 13 \text{ and } 5 + 11$$
$$18 = 5 + 13 \text{ and } 7 + 11$$
$$20 = 3 + 17 \text{ and } 7 + 13$$
$$50 = 19 + 31$$
$$100 = 53 + 47$$
$$21,000 = 17 + 20,983$$
$$.$$
$$.$$

Computer searches completed in 2000 had verified that all even numbers up to 400 trillion (4×10^{14}), which is not a small list, are sums of 2 primes, while in 2008, a distributed computer search ran by Tomas Oliveira e Silva, a researcher at the University of Aveiro, Portugal, had further verified the Goldbach conjecture up to 12×10^{17}. But is the conjecture valid?

Argument:

By Euclid's proof, there is an infinitude of primes; that is, the list of primes: 2, 3, 5, 7, 11, 13, 17, 19, 23, 29, 31, 37, … continues to infinity.

Goldbach's conjecture states that every even number after the number 2 is the sum of 2 primes. How do we prove this?

First of all, we ask a "reversed" question here (as opposed to Goldbach's conjecture). We ask whether all the prime numbers in the infinite list of prime numbers would combine with each other to form a regular, continuous (*without breaks or gaps*) and infinite list of even numbers. This would lead to our proof.

Let us now take a subset of primes from the infinite set of prime numbers, say, all the primes found in the set of integers ranging from 1 to 50, that is, the following subset of prime numbers:

2, 3, 5, 7, 11, 13, 17, 19, 23, 29, 31, 37, 41, 43 and 47

Then, we conduct a close examination of how this subset of prime numbers "behaves", that is, how the prime numbers combine (one-to-one) with each other to form even numbers, and observe whether there is any "regularity of pattern" in the way they do so. Here, we look at how the prime numbers from 2 To 47 combine with each other to form even numbers.

We could observe the primes from 2 to 47 "generating" a regular, continuous (*without breaks or gaps*) list of even numbers from 4 to 94. This is a regular, continuous list of even numbers, the even numbers becoming evidently progressively more repetitious. For example, there are 5 discernable combinations of primes/partitions for the even number 48, which is as follows:

a) $19 + 29 = 48$
b) $17 + 31 = 48$
c) $11 + 37 = 48$
d) $5 + 43 = 48$
e) $7 + 41 = 48$

and, there are 5 discernable combinations of primes/partitions for the even number 54, which is as follows:

a) $17 + 37 = 54$
b) $13 + 41 = 54$
c) $11 + 43 = 54$
d) $7 + 47 = 54$
e) $23 + 31 = 54$

And many others.

There appears to be a "regularity of pattern" in the way the even numbers "pop up". From this, we could thus conclude the following characteristic or "pattern" of the prime numbers: They would combine (one-to-one) with each other to form a regular, continuous (*without breaks or gaps*) list of even numbers, with "overwhelming repetitions" all over the place.

Next, we select another subset of primes from the infinite set of prime numbers. We would here take the prime numbers from the set of integers 51 to 100, which is just "next to" the set of integers 1 to 50 from which we have taken our first subset of prime numbers, to be our second subset of primes. This second subset of prime numbers is as follows:

53, 59, 61, 67, 71, 73, 79, 83, 89 and 97

As usual, we conduct a close examination of how this second subset of prime numbers "behaves," that is, how they combine (one-to-one) with each other to form even numbers, and observe whether there is "regularity of pattern" in the way they do so.

Here, as earlier, we could observe the primes from 53 to 97 "generating" a regular, continuous (*without breaks or gaps*) list of even numbers ranging from 56 to 194. This regular, continuous list of even numbers is also evidently progressively more repetitious. For example, there are 8 discernable combinations of primes/partitions for the even number 90, which is as follows:

a)   53 + 37 = 90
b)   59 + 31 = 90
c)   61 + 29 = 90
d)   67 + 23 = 90
e)   71 + 19 = 90
f)   73 + 17 = 90
g)   79 + 11 = 90
h)   83 + 7 = 90

and, there are 8 discernable combinations of primes/partitions for the even number 120, which is as follows:

a)   53 + 67 = 120
b)   59 + 61 = 120
c)   67 + 53 = 120
d)   73 + 47 = 120
e)   79 + 41 = 120
f)   83 + 37 = 120
g)   89 + 31 = 120
h)   97 + 23 = 120

And many others.

There appears to be a "regularity of pattern" in the way the even numbers "pop up" here - as a matter of fact, this "regularity of pattern" resembles that found in the earlier listing.

The list of even numbers "generated" by this second subset of prime numbers, that is, the regular, continuous list of even numbers ranging from 56 to 194, even overlaps (by a wide margin) the list of even numbers "generated" by the first subset of primes (2 to 47), that is, the regular, continuous list of even numbers ranging from 4 to 94.

From this listing also, from the "characteristics" found in all these listings, where the "regularity of pattern" of the appearance of the even numbers is evident, we could deduce the following characteristic of the prime numbers: The prime numbers would combine (one-to-one) with each other to form a regular, continuous (*without breaks or gaps*) list of even numbers, with "overwhelming repetitions" all over the place. This could be further confirmed by studying the even numbers "generated", e.g., by the 3 subsets of primes by combining with other prime numbers, including the prime numbers before them, for the 3 consecutive sets of integers, 101 To 150, 151 To 200, and, 201 To 250, with "overwhelming repetitions" all over the place (see Item (1) in the data below, where there are also much further examples).

Lemma: The well-established self-similarity concept, which was developed by Mitchell Feigenbaum in the 1970s and which brought him fame, upon which the method of renormalization in perturbation theory is based, postulates that there is a tendency of identical mathematical structures to recur on many levels. Within a given structure, there would be smaller copies of the same structure, their sizes being determined by the scaling factor. Feigenbaum found that at the utmost tips of the fig-tree, there is some mathematical structure which remains the same when its size is changed (enlarged) by a scaling factor of 4.669, which is found to be a constant like pi (3.142); this structure is the shape of the fig-tree itself; in other words, little whorls could be found within big whorls. Renormalization has been a well-established technique in chaos theory/fractal geometry and is a mathematical trick which functions rather like a microscope, zooming in on the self-similar structure, removing any approximations, and filtering out everything else. All this shows the universality of some features of chaos. That is, some kind of order or pattern could be found in or is inherent in disorder or chaos. In other words, the elements of an infinite subset of an infinite set contain all the recursive significant properties of that set unless the process which selects the elements of the subset directly excludes a property.

To make it simpler, we re-phrase this concept as follows: The characteristic of a mountain or infinite volume of sand is reflected in the characteristic of some grains of sand found there so that studying the characteristic of some grains of sand found there is sufficient for deducing the characteristic of the mountain or infinite volume of sand. Likewise, if x is a subset of y and if x is a list of prime numbers while y is another list of prime numbers, the characteristic presence of the even numbers "generated" by all the primes in x suggests (or reflects) the characteristic presence of even numbers "generated" by all the primes in y, so that if y is an infinite list of prime numbers, whence the prime numbers in it run to infinity, so do the even numbers "generated" by all the primes in it. What is described here is actually the "reflection" principle.

Therefore, by the above-mentioned principle, all the above-mentioned selected subsets of primes in the infinite set of primes would each reflect (present an image of, or, display something which has similarity with) the characteristic of this infinite set of primes; that is, all the infinite primes in the infinite set of primes, including its infinite subsets such as the selected subsets mentioned above, would combine (one-to-one) with each other to form a regular, continuous (*without breaks or gaps*) and infinite (implied by the infinitude of the primes (vide Euclid's proof) and the even numbers themselves) list of even numbers. It is evident that the higher up the infinite list of primes we go, the more "overwhelming" or dense the (one-to-one) combinations of primes (in the formation of even numbers) would become, the number of permutations of the combinations of primes tending towards infinity (with the infinity of the prime numbers), as a study with further selected subsets of primes would reveal. A study of the even numbers "generated" by all these subsets of primes would show that the higher up the subsets of primes we go the more "overwhelmingly" the even numbers "generated" would repeat themselves and overlap. This is (very) significant. Though the infinitude of the prime numbers would ensure that there would always be new even numbers being "generated", there is also the "fear" that there might be gaps, breaks or lack of continuity in the even numbers thus "generated". But, it is evident that these more and more profuse repetitions and overlaps of the even numbers thus "generated" by the primes the higher up the infinite list of prime numbers we go "ensure" that such gaps or breaks would not appear between the even numbers "generated" - they "ensure" that the even numbers thus "generated" by the primes in the infinite list of primes would be regular, continuous, *without breaks or gaps*, and, in consecutive running order. Also (very) significant is the great number of new even numbers that each of the primes in these subsets of primes helps to "generate". This "profuse generation" of "regular batches" of even numbers by the prime numbers represents a characteristic or feature of the prime numbers, a universal "pattern" or feature of the "chaotic" infinite prime numbers, or, recurrent, identical mathematical structures, which is all in accordance with the above lemma. (This "pattern of behaviour" of the prime numbers, as described in the chapter, is analogous to the "self-similar mathematical pattern or structure" (which is the shape of the fig-tree itself) of the various parts of the fig-tree, that is, its trunk to bough section, bough to branch section, branch to twig section and twig to twiglet section, in

Feigenbaum's famous fig-tree example, and, such self-similar mathematical pattern or structure, or, fractal characteristic, could also be found in other aspects of nature, for example, waves, turbulence or chaos, the structures of viruses and bacteria, polymers and ceramic materials, the universe and many others, even the movements of prices in financial markets, the growths of populations, the sound of music, the flow of blood through our circulatory system, the behaviour of people en masse, etc., which have all spawned a relatively new and important branch of mathematics with wide practical applications known as fractal geometry, which has been pioneered by Benoit Mandelbrot. As a matter of fact, self-similarity or fractal characteristic could be regarded as the fundamental mathematical aspect found in practically everything in nature including the numbers such as the prime numbers and the even numbers which are the subjects of our investigation here, and, this new branch of mathematics, fractal geometry, besides having a great practical impact on us also gives us a deeper vision of the universe in which we live and our place in it.) In other words, by the above lemma the infinity of the prime numbers implies the infinity of the "profuse generation" of "regular batches" of even numbers by the prime numbers, that is, the validity of the Goldbach conjecture.

Here, we take a close look at the following data:-

(1) No. Of Old/Repeated (Also Appeared Earlier) Even Numbers/Overlaps "Generated" (By The Additions/Combinations Of 2 Primes), For Integers 1 To 1,250 (*See Appendix 1 For Example Of Computation Method*)

(a) Set Of Integers, 1 To 50, With 14 Primes Within It = Not Applicable
(aa) Percentage Increase In Repetition = Not Applicable

(b) Set Of Integers, 51 To 100, With 10 Primes Within It = **20** Repeated Even Nos.
(bb) Percentage Increase In Repetition = Not Applicable

(c) Set Of Integers, 101 To 150, With 10 Primes Within It = 46 Repeated Even Nos.
(cc) Percentage Increase In Repetition = (46 - 20) ÷ 20 x 100% = **130%**

(d) Set Of Integers, 151 To 200, With 11 Primes Within It = 73 Repeated Even Nos.
(dd) Percentage Increase In Repetition = (73 - 46) ÷ 46 x 100% = 58.7%

(e) Set Of Integers, 201 To 250, With 7 Primes Within It = 93 Repeated Even Nos.
(ee) Percentage Increase In Repetition = (93 - 73) ÷ 73 x 100% = 27.4%

(f) Set Of Integers, 251 To 300, With 9 Primes Within It = 115 Repeated Even Nos.
(ff) Percentage Increase In Repetition = (115 - 93) ÷ 93 x 100% = 23.66%

(g) Set Of Integers, 301 To 350, With 8 Primes Within It = 139 Repeated Even Nos.
(gg) Percentage Increase In Repetition = (139 - 115) ÷ 115 x 100% = 20.87%

(h) Set Of Integers, 351 To 400, With 8 Primes Within It = 172 Repeated Even Nos.
(hh) Percentage Increase In Repetition = (172 - 139) ÷ 139 x 100% = 23.74%

(i) Set Of Integers, 401 To 450, With 9 Primes Within It = 196 Repeated Even Nos.
(ii) Percentage Increase In Repetition = (196 - 172) ÷ 172 x 100% = 13.95%

(j) Set Of Integers, 451 To 500, With 8 Primes Within It = 220 Repeated Even Nos.
(jj) Percentage Increase In Repetition = (220 - 196) ÷ 196 x 100% = 12.24%

(k) Set Of Integers, 501 To 550, With 6 Primes Within It = 247 Repeated Even Nos.
(kk) Percentage Increase In Repetition = (247 - 220) ÷ 220 x 100% = 12.27%

(l) Set Of Integers, 551 To 600, With 8 Primes Within It = 268 Repeated Even Nos.
(ll) Percentage Increase In Repetition = (268 - 247) ÷ 247 x 100% = 8.5%

(m) Set Of Integers, 601 To 650, With 9 Primes Within It = 298 Repeated Even Nos.
(mm) Percentage Increase In Repetition = (298 - 268) ÷ 268 x 100% = 11.19%

(n) Set Of Integers, 651 To 700, With 7 Primes Within It = 320 Repeated Even Nos.
(nn) Percentage Increase In Repetition = (320 - 298) ÷ 298 x 100% = 7.38%

(o) Set Of Integers, 701 To 750, With 7 Primes Within It = 340 Repeated Even Nos.
(oo) Percentage Increase In Repetition = (340 - 320) ÷ 320 x 100% = 6.25%

(p) Set Of Integers, 751 To 800, With 7 Primes Within It = 367 Repeated Even Nos.
(pp) Percentage Increase In Repetition = (367 - 340) ÷ 340 x 100% = 7.94%

(q) Set Of Integers, 801 To 850, With 7 Primes Within It = 392 Repeated Even Nos.
(qq) Percentage Increase In Repetition = (392 - 367) ÷ 367 x 100% = 6.81%

(r) Set Of Integers, 851 To 900, With 8 Primes Within It = 412 Repeated Even Nos.
(rr) Percentage Increase In Repetition = (412 - 392) ÷ 392 x 100% = 5.1%

(s) Set Of Integers, 901 To 950, With 7 Primes Within It = 433 Repeated Even Nos.
(ss) Percentage Increase In Repetition = (433 - 412) ÷ 412 x 100% = 5.1%

(t) Set Of Integers, 951 To 1,000, With 7 Primes Within It = 470 Repeated Even Nos.
(tt) Percentage Increase In Repetition = (470 - 433) ÷ 433 x 100% = 8.55%

(u) Set Of Integers, 1,001 To 1,050, With 8 Primes Within It = 492 Repeated Even Nos.
(uu) Percentage Increase In Repetition = (492 - 470) ÷ 470 x 100% = 4.68%

(v) Set Of Integers, 1,051 To 1,100, With 8 Primes Within It = 523 Repeated Even Nos.
(vv) Percentage Increase In Repetition = (523 - 492) ÷ 492 x 100% = 6.3%

(w) Set Of Integers, 1,101 To 1,150, With 5 Primes Within It = 545 Repeated Even Nos.
(ww) Percentage Increase In Repetition = (545 - 523) ÷ 523 x 100% = 4.21%

(x) Set Of Integers, 1,151 To 1,200, With 7 Primes Within It = 553 Repeated Even Nos.
(xx) Percentage Increase In Repetition = (553 - 545) ÷ 545 x 100% = **1.47%**

(y) Set Of Integers, 1,201 To 1,250, With 8 Primes Within It = **592** Repeated Even Nos.
(yy) Percentage Increase In Repetition = (592 - 553) ÷ 553 x 100% = **7.05%**

It could be seen that on the whole the No. Of Old/Repeated (Also Appeared Earlier) Even Numbers/Overlaps "Generated" (By The Additions/Combinations Of 2 Primes) increases progressively from 20 in (b) to 592 in (y), while it could be seen that the Percentage Increase In Repetition on the whole thins out from 130% in (cc) to 7.05% in (yy), with the lowest percentage increase of 1.47% found in (xx). This statistical trend or feature is not surprising and represents (very) significant evidence that lends support to the validity of the Goldbach conjecture - the infinitude of both the primes and the even numbers implies that the above overlaps increase progressively to infinity.

(2) Density Of New Even Numbers "Generated" (*See Appendix 1 For Example Of Computation Method*)

   (a) Set Of Integers, 51 To 100, With 10 Primes Within It = 5 New Even Nos. Per Prime No.
(No. Of New Even Nos. "Generated" = 50. No. Of Primes = 10.)

   (b) Set Of Integers, 101 To 150, With 10 Primes Within It = 5.2 New Even Nos. Per Prime No.
(No. Of New Even Nos. "Generated" = 52. No. Of Primes = 10.)

   (c) Set Of Integers, 151 To 200, With 11 Primes Within It = **4.55 New Even Nos. Per Prime No.**
(No. Of New Even Nos. "Generated" = 50. No. Of Primes = 11.)

   (d) Set Of Integers, 201 To 250, With 7 Primes Within It = 6 New Even Nos. Per Prime No.
(No. Of New Even Nos. "Generated" = 42. No. Of Primes = 7.)

   (e) Set Of Integers, 251 To 300, With 9 Primes Within It = 5.78 New Even Nos. Per Prime No.
(No. Of New Even Nos. "Generated" = 52. No. Of Primes = 9.)

   (f) Set Of Integers, 301 To 350, With 8 Primes Within It = 7 New Even Nos. Per Prime No.
(No. Of New Even Nos. "Generated" = 56. No. Of Primes = 8.)

   (g) Set Of Integers, 351 To 400, With 8 Primes Within It = 6 New Even Nos. Per Prime No.
(No. Of New Even Nos. "Generated" = 48. No. Of Primes = 8.)

   (h) Set Of Integers, 401 To 450, With 9 Primes Within It = 5.78 New Even Nos. Per Prime No.
(No. Of New Even Nos. "Generated" = 52. No. Of Primes = 9.)

   (i) Set Of Integers, 451 To 500, With 8 Primes Within It = 6.25 New Even Nos. Per Prime No.
(No. Of New Even Nos. "Generated" = 50. No. Of Primes = 8.)

   (j) Set Of Integers, 501 To 550, With 6 Primes Within It = 8 New Even Nos. Per Prime No.
(No. Of New Even Nos. "Generated" = 48. No. Of Primes = 6.)

   (k) Set Of Integers, 551 To 600, With 8 Primes Within It = 6.5 New Even Nos. Per Prime No.
(No. Of New Even Nos. "Generated" = 52. No. Of Primes = 8.)

   (l) Set Of Integers, 601 To 650, With 9 Primes Within It = 5.33 New Even Nos. Per Prime No.
(No. Of New Even Nos. "Generated" = 48. No. Of Primes = 9.)

   (m) Set Of Integers, 651 To 700, With 7 Primes Within It = 6.29 New Even Nos. Per Prime No.
(No. Of New Even Nos. "Generated" = 44. No. Of Primes = 7.)

   (n) Set Of Integers, 701 To 750, With 7 Primes Within It = 7.43 New Even Nos. Per Prime No.
(No. Of New Even Nos. "Generated" = 52. No. Of Primes = 7.)

   (o) Set Of Integers, 751 To 800, With 7 Primes Within It = 7.71 New Even Nos. Per Prime No.
(No. Of New Even Nos. "Generated" = 54. No. Of Primes = 7.)

   (p) Set Of Integers, 801 To 850, With 7 Primes Within It = 6 New Even Nos. Per Prime No.
(No. Of New Even Nos. "Generated" = 42. No. Of Primes = 7.)

   (q) Set Of Integers, 851 To 900, With 8 Primes Within It = 6 New Even Nos. Per Prime No.

(No. Of New Even Nos. "Generated" = 48. No. Of Primes = 8.)

(r) Set Of Integers, 901 To 950, With 7 Primes Within It = 8.57 New Even Nos. Per Prime No.
(No. Of New Even Nos. "Generated" = 60. No. Of Primes = 7.)

(s) Set Of Integers, 951 To 1,000, With 7 Primes Within It = 7.14 New Even Nos. Per Prime No.
(No. Of New Even Nos. "Generated" = 50. No. Of Primes = 7.)

(t) Set Of Integers, 1,001 To 1,050, With 8 Primes Within It = 6.5 New Even Nos. Per Prime No.
(No. Of New Even Nos. "Generated" = 52. No. Of Primes = 8.)

(u) Set Of Integers, 1,051 To 1,100, With 8 Primes Within It = 6 New Even Nos. Per Prime No.
(No. Of New Even Nos. "Generated" = 48. No. Of Primes = 8.)

(v) Set Of Integers, 1,101 To 1,150, With 5 Primes Within It = 6.4 New Even Nos. Per Prime No.
(No. Of New Even Nos. "Generated" = 32. No. Of Primes = 5.)

(w) Set Of Integers, 1,151 To 1,200, With 7 Primes Within It = **9.14 New Even Nos. Per Prime No.**
(No. Of New Even Nos. "Generated" = 64. No. Of Primes = 7.)

(x) Set Of Integers, 1,201 To 1,250, With 8 Primes Within It = 7 New Even Nos. Per Prime No.
(No. Of New Even Nos. "Generated" = 56. No. Of Primes = 8.)

Average Density For The Above 24 Items ((a) To (x)) = 155.54 ÷ 24 = 6.48 New Even Nos. Per Prime No.

Maximum Density = 9.14 New Even Nos. Per Prime No. (No. Of New Even Nos. "Generated" = 64. No. Of Primes = 7.)

Minimum Density = 4.55 New Even Nos. Per Prime No. (No. Of New Even Nos. "Generated" = 50. No. Of Primes = 11.)

Such a "profuse generation" of "regular batches" of even numbers by the prime numbers is (very) significant and represents a characteristic or feature of the prime numbers, a universal "pattern" or feature of the "chaotic" infinite prime numbers (or, recurrent, identical mathematical structures), which is excellently in accordance with the above lemma. This lends further support to the validity of the Goldbach conjecture, which, as stated above, is implied by both the infinitude of the primes and the even numbers. There is indeed further incontrovertible evidence which is obtainable by analysing a number of even numbers; e.g., we could split a group of 240 even consecutive numbers, from 4 to 482, into 8 equal batches (30 even numbers per batch) and analyse the batches, which buttresses the above evidence that the infinite quantity of primes would "generate" a regular, continuous (*without breaks or gaps*) and infinite list of even numbers. The density of distribution or prime additions/combinations per even number evidently become greater and greater the higher up the infinite list of the even numbers we go - this increase in density evidently represents a definite pattern in the "behaviour" of the prime numbers. This pattern is (very) significant and is discernable in the following example:-

(1) <u>1 st. Batch Of 30 Even Numbers (4 To 62)</u>   (*See Appendix 2 For Example Of Computation Method*)

   a) Maximum No. Of Prime Additions/Combinations Per Even Number = **5**
   b) Minimum No. Of Prime Additions/Combinations Per Even Number = **1**
   c) Density Of Distribution = Average Prime Additions/Combinations Per Even Number = **2.77** Prime Additions/Combinations Per Even Number (see Appendix for computation method)

(2) 2 nd. Batch Of 30 Even Numbers (64 To 122)

    a) Maximum No. Of Prime Additions/Combinations Per Even Number = 14
    b) Minimum No. Of Prime Additions/Combinations Per Even Number = 2
    c) Density Of Distribution = Average Prime Additions/Combinations Per Even Number = **6.1** Prime Additions/Combinations Per Even Number
    d) Percentage Increase In Density Of Distribution = (6.1 - 2.77) ÷ 2.77 x 100% = 120.22%

(3) 3 rd. Batch Of 30 Even Numbers (124 To 182)

    a) Maximum No. Of Prime Additions/Combinations Per Even Number = 16
    b) Minimum No. Of Prime Additions/Combinations Per Even Number = 4
    c) Density Of Distribution = Average Prime Additions/Combinations Per Even Number = **9.07** Prime Additions/Combinations Per Even Number
    d) Percentage Increase In Density Of Distribution = (9.07 - 6.1) ÷ 6.1 x 100% = 48.69%

(4) 4 th. Batch Of 30 Even Numbers (184 To 242)

    a) Maximum No. Of Prime Additions/Combinations Per Even Number = 22
    b) Minimum No. Of Prime Additions/Combinations Per Even Number = 5
    c) Density Of Distribution = Average Prime Additions/Combinations Per Even Number = **10.53** Prime Additions/Combinations Per Even Number
    d) Percentage Increase In Density Of Distribution = (10.53 - 9.07) ÷ 9.07 x 100% = 16.1%

(5) 5 th. Batch Of 30 Even Numbers (244 To 302)

    a) Maximum No. Of Prime Additions/Combinations Per Even Number = 21
    b) Minimum No. Of Prime Additions/Combinations Per Even Number = 7
    c) Density Of Distribution = Average Prime Additions/Combinations Per Even Number = **12.37** Prime Additions/Combinations Per Even Number
    d) Percentage Increase In Density Of Distribution = (12.37 - 10.53) ÷ 10.53 x 100% = 17.47%

(6) 6 th. Batch Of 30 Even Numbers (304 To 362)

    a) Maximum No. Of Prime Additions/Combinations Per Even Number = 27
    b) Minimum No. Of Prime Additions/Combinations Per Even Number = 7
    c) Density Of Distribution = Average Prime Additions/Combinations Per Even Number = **13.77** Prime Additions/Combinations Per Even Number
    d) Percentage Increase In Density Of Distribution = (13.77 - 12.37) ÷ 12.37 x 100% = 11.32%

(7) 7 th. Batch Of 30 Even Numbers (364 To 422)

    a) Maximum No. Of Prime Additions/Combinations Per Even Number = 30
    b) Minimum No. Of Prime Additions/Combinations Per Even Number = 7
    c) Density Of Distribution = Average Prime Additions/Combinations Per Even Number = **15.23** Prime Additions/Combinations Per Even Number
    d) Percentage Increase In Density Of Distribution = (15.23 - 13.77) ÷ 13.77 x 100% = 10.6%

(8) <u>8 th. Batch Of 30 Even Numbers (424 To 482)</u>

    a)   Maximum No. Of Prime Additions/Combinations Per Even Number = **30**
    b)   Minimum No. Of Prime Additions/Combinations Per Even Number = **9**
    c)   Density Of Distribution = Average Prime Additions/Combinations Per Even Number = **16.93**
        Prime Additions/Combinations Per Even Number
    d)   Percentage Increase In Density Of Distribution = (16.93 - 15.23) ÷ 15.23 x 100% = 11.16%

The Density Of Distribution is expected to increase to infinity, though the Percentage Increase In Density Of Distribution is expected to thin out towards infinity - it could be seen above to increase from 2.77 prime additions/combinations per even number for batch of even numbers, 4 to 62, all the way up to 16.93 prime additions/combinations per even number for batch of even numbers, 424 to 482. This is nevertheless (very) significant evidence that lends support to the validity of the Goldbach conjecture. Also, the Maximum No. Of Prime Additions/Combinations Per Even Number and the Minimum No. Of Prime Additions/Combinations Per Even Number could be seen to range from 5 and 1 respectively for batch of even numbers, 4 to 62, to 30 and 9 respectively for batch of even numbers, 424 to 482. This trend of "upward increase" of the (maximum and minimum) numbers of prime additions/combinations for each even number implies that at some points toward infinity the numbers of prime additions/combinations for each even number could be thousands, millions, billions, trillions, and more, if only we have the computing power to compute/check such prime additions/combinations. This is (very) significant too and is also evidence that lends support to the validity of the Goldbach conjecture. By the infinitude of the primes and even numbers and the above lemma, these "patterns", as described here, would be there all the way to infinity, which would be in accordance with the Goldbach conjecture.

The one-to-one additions/combinations of the primes in the formation of even numbers do evidently become more and more "overwhelming" or profuse the higher up the infinite list of even numbers/prime numbers we go, thereby assuring an infinite, regular and consecutive supply of even numbers, as is evident from the example just above. This, together with the above-described evidently more and more profuse repetitions and overlaps of the even numbers "generated" by the primes the higher up the infinite list of prime numbers we go (refer to No. Of Old/Repeated (Also Appeared Earlier) Even Numbers/Overlaps "Generated" (By The Additions/Combinations Of 2 Primes), For Integers 1 To 1,250 above), go to show that the Goldbach conjecture becomes, evidently, even stronger and stronger the higher up the infinite list of prime numbers/even numbers we go. Here, we have in fact approached the problem from 2 different, but somewhat related, angles - by a statistical analysis of the "behaviour" of the primes in the formation of even numbers, and, a statistical analysis of the even numbers "generated" as a result. The statistical data thus obtained are indeed found to greatly support the Goldbach conjecture, evidently the more so the higher up the infinite list of prime numbers/even numbers we go, and, by the infinitude of the primes and even numbers and the above lemma there would be an infinitude of such statistical data thus obtained. Hence, by virtue of these imposing statistical trends, plus the statistical trend that a prolific number of new even numbers are always being "generated" (refer to Density Of New Even Numbers "Generated" above), as well as the infinitude of the prime numbers and the even numbers, together with the above lemma, we affirm the validity of the Goldbach conjecture.

Reversing the "reversed way", we hereby affirm that every even number after the number 2 in the infinite list of even numbers is a combination or sum of 2 primes. In fact, the prime numbers are the building-blocks or "atoms" of all the even numbers - and more - the prime numbers are the building-blocks of all the integers or whole numbers: every even number (with the exception of 2) is the sum of 2 prime numbers, and, every odd number is either a prime number, or, a composite of prime numbers (that is, the odd number has prime factors). It is truly the peculiar characteristics of the prime numbers themselves (as described above, whose distribution could in fact be predicted by the prime number theorem which had been proven, implying some pattern or fractal nature in the prime numbers as per the above lemma), which could be regarded as a self-similar or fractal feature as such, that are responsible for the Goldbach conjecture being true. By induction the Goldbach conjecture has been proven true - the above constitutes proof of the Goldbach conjecture (which ought to be known as the Goldbach Theorem instead).

This proof could be extended here. It has been mentioned above that the Goldbach conjecture had been tested and found to be correct for every even number up to 12×10^{17} by computer searches completed in 2008. Thus, by the above lemma, and, the infinitude of the primes and even numbers, this long list of consecutive even numbers up to 12×10^{17} reflects (indicates or implies) the fact that all the infinite even numbers above 12×10^{17} would each be the sum of 2 primes. (This is in accordance with the "reflection" principle stated above, wherein it is mentioned that the characteristic of a mountain or infinite volume of sand is reflected in the characteristic of some grains of sand found there so that studying the characteristic of some grains of sand found there is enough for deducing the characteristic of the mountain or infinite volume of sand.)

We therefore declare that the Goldbach conjecture is true - every even number after the number 2 is indeed the sum of 2 primes.

APPENDIX 1

(20) <u>Set Of Integers, 1,201 To 1,250, With 8 Primes Within It</u>
    (a) Primes: 1,201; 1,213; 1,217; 1,223; 1,229; 1,231; 1,237 and 1,249
    (b) No. Of Primes: 8
    (c) No. Of Even Numbers "Generated" (Including Repetitions) By The 8 Primes = 648 (1,204 [1,201 + 3] To
       2,498 [1,249 + 1,249])
    (d) No. Of New Even Numbers "Generated" = 56 (2,388 To 2,498)
    (e) No. Of Old/Repeated (Also Appeared In (19) Above, With Some Also Having Appeared In (18), (17), (16),
       (15), (14), (13), (12), (11), (10), (9) And (8) Above) Even Numbers "Generated" = 592 (1,204 To 2,386)
    (f) Density Of New Even Numbers "Generated" = (d) ÷ 8 Primes = 56 ÷ 8 = 7 New Even Numbers Per Prime
       Number

APPENDIX 2

(8) <u>8 th. Batch Of 30 Even Numbers (424 To 482)</u>
    (a) 424: No. Of Above-mentioned Prime Additions/Combinations = 12
    (b) 426: No. Of Above-mentioned Prime Additions/Combinations = 21
    (c) 428: No. Of Above-mentioned Prime Additions/Combinations = **9**
    (d) 430: No. Of Above-mentioned Prime Additions/Combinations = 14
    (e) 432: No. Of Above-mentioned Prime Additions/Combinations = 19
    (f) 434: No. Of Above-mentioned Prime Additions/Combinations = 14
    (g) 436: No. Of Above-mentioned Prime Additions/Combinations = 11
    (h) 438: No. Of Above-mentioned Prime Additions/Combinations = 22
    (i) 440: No. Of Above-mentioned Prime Additions/Combinations = 15
    (j) 442: No. Of Above-mentioned Prime Additions/Combinations = 13
    (k) 444: No. Of Above-mentioned Prime Additions/Combinations = 22
    (l) 446: No. Of Above-mentioned Prime Additions/Combinations = 12
    (m) 448: No. Of Above-mentioned Prime Additions/Combinations = 13
    (n) 450: No. Of Above-mentioned Prime Additions/Combinations = 29
    (o) 452: No. Of Above-mentioned Prime Additions/Combinations = 14
    (p) 454: No. Of Above-mentioned Prime Additions/Combinations = 12
    (q) 456: No. Of Above-mentioned Prime Additions/Combinations = 26
    (r) 458: No. Of Above-mentioned Prime Additions/Combinations = **9**
    (s) 460: No. Of Above-mentioned Prime Additions/Combinations = 17
    (t) 462: No. Of Above-mentioned Prime Additions/Combinations = **30**
    (u) 464: No. Of Above-mentioned Prime Additions/Combinations = 13
    (v) 466: No. Of Above-mentioned Prime Additions/Combinations = 14
    (w) 468: No. Of Above-mentioned Prime Additions/Combinations = 26

  (x) 470: No. Of Above-mentioned Prime Additions/Combinations = 16
  (y) 472: No. Of Above-mentioned Prime Additions/Combinations = 14
  (z) 474: No. Of Above-mentioned Prime Additions/Combinations = 24
 (aa) 476: No. Of Above-mentioned Prime Additions/Combinations = 14
 (bb) 478: No. Of Above-mentioned Prime Additions/Combinations = 12
 (cc) 480: No. Of Above-mentioned Prime Additions/Combinations = **30**
 (dd) 482: No. Of Above-mentioned Prime Additions/Combinations = 11
   (i) Maximum No. Of Prime Additions/Combinations = 30
  (ii) Minimum No. Of Prime Additions/Combinations = 9
 (iii) Total No. Of Prime Additions/Combinations For (a) To (dd) = 508
 (iv) Total No. Of Even Numbers = 30
  (v) Density Of Distribution = Average Prime Additions/Combinations Per Even Number = (iii) ÷ (iv) = 508
     ÷ 30 = 16.93 Prime Additions/Combinations Per Even Number

PART 2

Theorem:- Every even number after 2 is the sum of 2 primes.

Argument 1:-
Lemma: By Euclid's proof the primes are infinite.

The prime number theorem, which had been proven, states that the limit of the quotient of the 2 functions $\pi(n)$ and $n/\log n$ as n approaches infinity is 1, which is expressed by the formula:

$$\lim_{n \to \infty} \pi(n)/(n/\log n) = 1, \quad \text{where } \pi(n) \text{ is approximately equal to } (n/\log n)$$

The function $\pi(n)$ represents the number of primes less than or equal to the number n. This function measures the distribution of the prime numbers. With it, we compute the ratio $n/\pi(n)$ which says what fraction of the numbers up to a given point are primes. (It is actually the reciprocal of this fraction.) The following is the result of a computation:-

n	$\pi(n)$	$n/\pi(n)$
10	4 (a)	2.5
100	25 (b)	4.0
1,000	168 (c)	6.0
10,000	1,229 (d)	8.1
100,000	9,592 (e)	10.4
1,000,000	78,498 (f)	12.7
10,000,000	664,579 (g)	15.0
100,000,000	5,761,455 (h)	17.4
1,000,000,000	50,847,534 (i)	19.7
10,000,000,000	455,052,512 (j)	22.0

It is noticeable that as one moves from 1 power of 10 to the next, the ratio $n/\pi(n)$ increases by about 2.3, e.g., 22.0 - 19.7 = 2.3. As $\log_e 10 = 2.30258 \ldots$, we may thus regard $\pi(n)$ as approximately equal to $n/\log n$.

We have the following partitions with the primes described in the "$\pi(n)$" column above:-

1) With (a) above, we have the following "prime + prime = even number" combinations:

    a) prime a + prime a: 4 x 4 "prime + prime" combinations
    b) prime a + prime b: 4 x 25 "prime + prime" combinations
    c) prime a + prime c: 4 x 168 "prime + prime" combinations
    d) prime a + prime d: 4 x 1,229 "prime + prime" combinations
    e) prime a + prime e: 4 x 9,592 "prime + prime" combinations
    f) prime a + prime f: 4 x 78,498 "prime + prime" combinations
    g) prime a + prime g: 4 x 664,579 "prime + prime" combinations
    h) prime a + prime h: 4 x 5,761,455 "prime + prime" combinations
    i) prime a + prime i: 4 x 50,847,534 "prime + prime" combinations
    j) prime a + prime j: 4 x 455,052,512 "prime + prime" combinations

For example, for (j) above, a prime described in (a) in the "$\pi(n)$" column above plus a prime described in (j) in the "$\pi(n)$" column above give an even number, and there are 4 x 455,052,512 such "prime + prime = even number" combinations.

2) With (b) above, we have the following "prime + prime = even number" combinations:

    a) prime b + prime a: 25 x 4 "prime + prime" combinations
    b) prime b + prime b: 25 x 25 "prime + prime" combinations
    c) prime b + prime c: 25 x 168 "prime + prime" combinations
    d) prime b + prime d: 25 x 1,229 "prime + prime" combinations
    e) prime b + prime e: 25 x 9,592 "prime + prime" combinations
    f) prime b + prime f: 25 x 78,498 "prime + prime" combinations
    g) prime b + prime g: 25 x 664,579 "prime + prime" combinations
    h) prime b + prime h: 25 x 5,761,455 "prime + prime" combinations
    i) prime b + prime i: 25 x 50,847,534 "prime + prime" combinations
    j) prime b + prime j: 25 x 455,052,512 "prime + prime" combinations

3) With (c) above, we have the following "prime + prime = even number" combinations:

    a) prime c + prime a: 168 x 4 "prime + prime" combinations
    b) prime c + prime b: 168 x 25 "prime + prime" combinations
    c) prime c + prime c: 168 x 168 "prime + prime" combinations
    d) prime c + prime d: 168 x 1,229 "prime + prime" combinations
    e) prime c + prime e: 168 x 9,592 "prime + prime" combinations
    f) prime c + prime f: 168 x 78,498 "prime + prime" combinations
    g) prime c + prime g: 168 x 664,579 "prime + prime" combinations
    h) prime c + prime h: 168 x 5,761,455 "prime + prime" combinations
    i) prime c + prime i: 168 x 50,847,534 "prime + prime" combinations
    j) prime c + prime j: 168 x 455,052,512 "prime + prime" combinations

4) With (d) above, we have the following "prime + prime = even number" combinations:

a) prime d + prime a: 1,229 x 4 "prime + prime" combinations
b) prime d + prime b: 1,229 x 25 "prime + prime" combinations
c) prime d + prime c: 1,229 x 168 "prime + prime" combinations
d) prime d + prime d: 1,229 x 1,229 "prime + prime" combinations
e) prime d + prime e: 1,229 x 9,592 "prime + prime" combinations
f) prime d + prime f: 1,229 x 78,498 "prime + prime" combinations
g) prime d + prime g: 1,229 x 664,579 "prime + prime" combinations
h) prime d + prime h: 1,229 x 5,761,455 "prime + prime" combinations
i) prime d + prime i: 1,229 x 50,847,534 "prime + prime" combinations
j) prime d + prime j: 1,229 x 455,052,512 "prime + prime" combinations

5) With (e) above, we have the following "prime + prime = even number" combinations:

a) prime e + prime a: 9,592 x 4 "prime + prime" combinations
b) prime e + prime b: 9,592 x 25 "prime + prime" combinations
c) prime e + prime c: 9,592 x 168 "prime + prime" combinations
d) prime e + prime d: 9,592 x 1,229 "prime + prime" combinations
e) prime e + prime e: 9,592 x 9,592 "prime + prime" combinations
f) prime e + prime f: 9,592 x 78,498 "prime + prime" combinations
g) prime e + prime g: 9,592 x 664,579 "prime + prime" combinations
h) prime e + prime h: 9,592 x 5,761,455 "prime + prime" combinations
i) prime e + prime i: 9,592 x 50,847,534 "prime + prime" combinations
j) prime e + prime j: 9,592 x 455,052,512 "prime + prime" combinations

6) With (f) above, we have the following "prime + prime = even number" combinations:

a) prime f + prime a: 78,498 x 4 "prime + prime" combinations
b) prime f + prime b: 78,498 x 25 "prime + prime" combinations
c) prime f + prime c: 78,498 x 168 "prime + prime" combinations
d) prime f + prime d: 78,498 x 1,229 "prime + prime" combinations
e) prime f + prime e: 78,498 x 9,592 "prime + prime" combinations
f) prime f + prime f: 78,498 x 78,498 "prime + prime" combinations
g) prime f + prime g: 78,498 x 664,579 "prime + prime" combinations
h) prime f + prime h: 78,498 x 5,761,455 "prime + prime" combinations
i) prime f + prime i: 78,498 x 50,847,534 "prime + prime" combinations
j) prime f + prime j: 78,498 x 455,052,512 "prime + prime" combinations

7) With (g) above, we have the following "prime + prime = even number" combinations:

a) prime g + prime a: 664,579 x 4 "prime + prime" combinations
b) prime g + prime b: 664,579 x 25 "prime + prime" combinations

c) prime g + prime c: 664,579 x 168 "prime + prime" combinations
d) prime g + prime d: 664,579 x 1,229 "prime + prime" combinations
e) prime g + prime e: 664,579 x 9,592 "prime + prime" combinations
f) prime g + prime f: 664,579 x 78,498 "prime + prime" combinations
g) prime g + prime g: 664,579 x 664,579 "prime + prime" combinations
h) prime g + prime h: 664,579 x 5,761,455 "prime + prime" combinations
i) prime g + prime i: 664,579 x 50,847,534 "prime + prime" combinations
j) prime g + prime j: 664,579 x 455,052,512 "prime + prime" combinations

8) With (h) above, we have the following "prime + prime = even number" combinations:

a) prime h + prime a: 5,761,455 x 4 "prime + prime" combinations
b) prime h + prime b: 5,761,455 x 25 "prime + prime" combinations
c) prime h + prime c: 5,761,455 x 168 "prime + prime" combinations
d) prime h + prime d: 5,761,455 x 1,229 "prime + prime" combinations
e) prime h + prime e: 5,761,455 x 9,592 "prime + prime" combinations
f) prime h + prime f: 5,761,455 x 78,498 "prime + prime" combinations
g) prime h + prime g: 5,761,455 x 664,579 "prime + prime" combinations
h) prime h + prime h: 5,761,455 x 5,761,455 "prime + prime" combinations
i) prime h + prime i: 5,761,455 x 50,847,534 "prime + prime" combinations
j) prime h + prime j: 5,761,455 x 455,052,512 "prime + prime" combinations

9) With (i) above, we have the following "prime + prime = even number" combinations:

a) prime i + prime a: 50,847,534 x 4 "prime + prime" combinations
b) prime i + prime b: 50,847,534 x 25 "prime + prime" combinations
c) prime i + prime c: 50,847,534 x 168 "prime + prime" combinations
d) prime i + prime d: 50,847,534 x 1,229 "prime + prime" combinations
e) prime i + prime e: 50,847,534 x 9,592 "prime + prime" combinations
f) prime i + prime f: 50,847,534 x 78,498 "prime + prime" combinations
g) prime i + prime g: 50,847,534 x 664,579 "prime + prime" combinations
h) prime i + prime h: 50,847,534 x 5,761,455 "prime + prime" combinations
i) prime i + prime i: 50,847,534 x 50,847,534 "prime + prime" combinations
j) prime i + prime j: 50,847,534 x 455,052,512 "prime + prime" combinations

10) With (j) above, we have the following "prime + prime = even number" combinations:

a) prime j + prime a: 455,052,512 x 4 "prime + prime" combinations
b) prime j + prime b: 455,052,512 x 25 "prime + prime" combinations
c) prime j + prime c: 455,052,512 x 168 "prime + prime" combinations
d) prime j + prime d: 455,052,512 x 1,229 "prime + prime" combinations
e) prime j + prime e: 455,052,512 x 9,592 "prime + prime" combinations
f) prime j + prime f: 455,052,512 x 78,498 "prime + prime" combinations
g) prime j + prime g: 455,052,512 x 664,579 "prime + prime" combinations

h) prime j + prime h: 455,052,512 x 5,761,455 "prime + prime" combinations
i) prime j + prime i: 455,052,512 x 50,847,534 "prime + prime" combinations
j) prime j + prime j: 455,052,512 x 455,052,512 "prime + prime" combinations

.

.

.

The above partitions/"prime + prime = even number" combinations are evidently progressively more "overwhelming" and repetitive. It is not surprising that computer searches completed in 2000 had verified that all even numbers up to 400 trillion (4×10^{14}), which is not a small list, are sums of 2 primes, while in 2008, a distributed computer search ran by Tomas Oliveira e Silva, a researcher at the University of Aveiro, Portugal, had further verified the Goldbach conjecture up to 12×10^{17}.

The infinitude of the primes, as per the above lemma, together with the infinitude of the even numbers, however imply that the above partitions/"prime + prime = even number" combinations would become increasingly more "overwhelming", dense, and repetitive towards infinity (the Goldbach conjecture becoming evidently stronger and stronger the higher up the infinite list of prime numbers/even numbers we go), hence "ensuring" the continuity (without any breaks or gaps) of the even numbers, and would be so all the way to infinity, thus proving that every even number after 2 is the sum of 2 primes.

Argument 2:-
Lemma: According to the principle of complete induction in set theory, if a set of natural numbers contains 1 and, for each n, it contains $n + 1$ whenever it contains all numbers less than $n + 1$, then it must contain every natural number, e.g., complete induction proves that every natural number is a product of primes.

By the above lemma, every even number after 2 in the infinite set of the integers is the sum of 2 primes; as per the distributed computer search completed in 2008 at the University of Aveiro, Portugal, stated above, 12×10^{17} in the infinite set of the integers is the largest even number found to be the sum of 2 primes while all the consecutive even numbers before it, from 4 to $(12 \times 10^{17}) - 2$, which is not a small list of numbers (it is in fact a long, impressive list, obtainable only with the help of modern computer technology), are also found to be sums of 2 primes - the principle of complete induction implies that all even numbers after 12×10^{17} in the infinite set of the integers must also be sums of 2 primes, i.e., it implies that every even number after 2 in the infinite set of the integers must be the sum of 2 primes - in other words, the Goldbach conjecture must be true.

Argument 3:-
Lemma: By Euclid's proof the primes are infinite.

We make use of the argument by "reductio ad absurdum" here. For this indirect argument, we assume that the Goldbach conjecture is false. (Before we proceed further, we should again note that a long, impressive list of consecutive even numbers, from 4 to 12×10^{17}, had already been verified to be sums of 2 primes, and, these partitions/"prime + prime = even number" combinations would become increasingly more "overwhelming", dense, and repetitive towards infinity (the Goldbach conjecture becoming evidently stronger and stronger the higher up the infinite list of prime numbers/even numbers we go), as is described above. The moot question now is, of course, whether after 12×10^{17} there would be an even number in the infinite list of even numbers which is the last, or, largest, even number which is the sum of 2 primes - this largest even number, if it exists (thereby proving the falsehood of the Goldbach conjecture), must (of necessity) be the sum of 2 primes which are each the largest existing prime. Before we continue, this point should be clearly held in mind.) This assumption implies that there is a limit to the even numbers which are sums of 2 primes and that there is a largest even number (e) which is, and must necessarily be, the sum of 2 primes that are each the largest existing prime ($e = x + x$, this largest even number, e, representing the ultimate limit of the even numbers which are sums of 2 primes, the 2 primes which add up to give e being of necessity each the largest existing prime (x)). This is of course a contradiction of the above lemma, which would imply that the lemma is false. But the lemma cannot be false - it is in fact a theorem (which had been proven by Euclid); there cannot be a largest existing prime (x) - the primes are infinite. This means that our assumption that the Goldbach conjecture is false is untenable and that the Goldbach conjecture must be true, i.e., every even number after 2 must be the sum of 2 primes. As a matter of fact, the above lemma implies that there would be an infinite number of double primes which sum up to an even number.

By both induction and contradiction the Goldbach conjecture is hence proved.

PART 3

Theorem:- Every even number after 2 is the sum of 2 primes.

Argument 1:-

Every even number after 2 is the sum of 2 odd numbers. Every odd number is either a prime which is odd or a composite - product of primes which are odd; notably, every prime with the exception of 2 is an odd number. Every even number after 2 is also a composite, but, a composite with at least 1 even prime factor, namely, 2, while the rest of its prime factors are odd, i.e., it is an even composite.

Therefore, every even number after 2 is the sum of 2 primes which are odd and/or the sum of 1 prime which is odd and 1 odd composite whose prime factors are odd and/or the sum of 2 odd composites whose prime factors are odd, besides being an even composite with at least 1 even prime factor, namely, 2, while the rest of its prime factors are odd.

Lemma:

By Euclid's proof, the primes are infinite; this implies that there would be an infinitude of sums of 2 primes as per the Goldbach conjecture. The even numbers, which are sums of 2 primes as per the conjecture, are also infinite. Thus, there are an infinite number of even numbers which are sums of 2 primes, both the even numbers and sums of 2 primes being infinite.

Corollary:

The odd numbers, which are either prime, every prime with the exception of 2 being an odd number, or composite (have prime factors which are odd), are infinite; this implies that there would be an infinite number of sums of 2 odd numbers, each of which is equal to an even number. Hence, as there is an infinitude of even numbers which are sums of 2 primes, as per the above lemma, and as all primes with the exception of

2 are odd numbers, there are an infinite number of even numbers which are sums of 2 odd numbers that are prime, all the even numbers, sums of 2 odd numbers and primes being infinite; i.e., every even number after 2 is also the sum of 2 odd numbers that are prime.

We thereby see the close interlink or relationship between the primes, even numbers and odd numbers, which are all infinite, which is significant.

The following are thus evident:

a) Every sum of 2 primes which are odd numbers is equal to an even number, as is below in consecutive order:

$2 + 2 = 1 + 3 = 4$

$3 + 3 = 1 + 5 = 6$

$3 + 5 = 1 + 7 = 8$

$5 + 5 = 3 + 7 = 10$

$5 + 7 = 1 + 11 = 12$

$7 + 7 = 3 + 11 = 1 + 13 = 14$

$3 + 13 = 5 + 11 = 16$

$7 + 11 = 5 + 13 = 1 + 17 = 18$

$7 + 13 = 3 + 17 = 1 + 19 = 20$

$11 + 11 = 3 + 19 = 5 + 17 = 11 + 11 = 22$

$11 + 13 = 5 + 19 = 7 + 17 = 1 + 23 = 24$

$13 + 13 = 3 + 23 = 7 + 19 = 26$

$11 + 17 = 5 + 23 = 28$

$13 + 17 = 11 + 19 = 7 + 23 = 1 + 29 = 30$

$3 + 29 = 13 + 19 = 1 + 31 = 32$

17 + 17 = 3 + 31 = 5 + 29 = 11 + 23 = 17 + 17 = **34**

17 + 19 = 5 + 31 = 7 + 29 = 13 + 23 = **36**

19 + 19 = 7 + 31 = 1 + 37 = **38**

3 + 37 = 11 + 29 = 17 + 23 = **40**

19 + 23 = 5 + 37 = 11 + 31 = 13 + 29 = 1 + 41 = **42**

3 + 41 = 7 + 37 = 13 + 31 = 1 + 43 = **44**

23 + 23 = 3 + 43 = 5 + 41 = 17 + 29 = **46**

5 + 43 = 7 + 41 = 11 + 37 = 17 + 31 = 19 + 29 = 1 + 47 = **48**

3 + 47 = 7 + 43 = 13 + 37 = 19 + 31 = **50**

23 + 29 = 5 + 47 = 11 + 41 = **52**

7 + 47 = 11 + 43 = 13 + 41 = 17 + 37 = 23 + 31 = 1 + 53 = **54**

3 + 53 = 13 + 43 = 19 + 37 = **56**

29 + 29 = 5 + 53 = 11 + 47 = 17 + 41 = 29 + 29 = **58**

29 + 31 = 7 + 53 = 13 + 47 = 17 + 43 = 19 + 41 = 23 + 37 = 1 + 59 = **60**

31 + 31 = 3 + 59 = 19 + 43 = 1 + 61 = **62**

3 + 61 = 5 + 59 = 11 + 53 = 17 + 47 = 23 + 41 = **64**

5 + 61 = 7 + 59 = 13 + 53 = 19 + 47 = 23 + 43 = 29 + 37 = **66**

7 + 61 = 31 + 37 = 1 + 67 = **68**

3 + 67 = 11 + 59 = 17 + 53 = 23 + 47 = 29 + 41 = **70**

5 + 67 = 11 + 61 = 13 + 59 = 19 + 53 = 29 + 43 = 31 + 41 = 1 + 71 = **72**

37 + 37 = 3 + 71 = 7 + 67 = 13 + 61 = 31 + 43 = 37 + 37 = 1 + 73 = **74**

3 + 73 = 5 + 71 = 17 + 59 = 23 + 53 = 29 + 47 = **76**

37 + 41 = 5 + 73 = 7 + 71 = 11 + 67 = 31 + 47 = 37 + 41 = **78**

7 + 73 = 13 + 67 = 19 + 61 = 37 + 43 = 1 + 79 = **80**

41 + 41 = 3 + 79 = 11 + 71 = 23 + 59 = 29 + 53 = **82**

41 + 43 = 5 + 79 = 11 + 73 = 13 + 71 = 17 + 67 = 23 + 61 = 31 + 53 = 37 + 47 = 1+ 83 = **84**

43 + 43 = 3 + 83 = 7 + 79 = 13 + 73 = 19 + 67 = 43 + 43 = **86**

5 + 83 = 17 + 71 = 29 + 59 = 41 + 47 = **88**

7 + 83 = 11 + 79 = 17 + 73 = 19 + 71 = 23 + 67 = 29 + 61 = 31 + 59 = 37 + 53 = 43 + 47 = 1 + 89 = **90**

3 + 89 = 13 + 79 = 19 + 73 = 31 + 61 = 1 + 91 = **92**

47 + 47 = 5 + 89 = 11 + 83 = 23 + 71 = 41 + 53 = 47 + 47 = **94**

5 + 91 = 7 + 89 = 13 + 83 = 17 + 79 = 23 + 73 = 29 + 67 = 37 + 59 = 43 + 53 = **96**

7 + 91 = 19 + 79 = 31 + 67 = 37 + 61 = 1 + 97 = **98**

47 + 53 = 3 + 97 = 11 + 89 = 17 + 83 = 29 + 71 = 41 + 59 = 47 + 53 = **100**

5 + 97 = 11 + 91 = 13 + 89 = 19 + 83 = 23 + 79 = 29 + 73 = 31 + 71 = 41+ 61 = 43 + 59 = 1 + 101 = **102**

.

.

.

b) Every sum of 1 prime which is an odd number & 1 odd composite which is the product of primes which are odd, is equal to the sum of 2 primes which are odd numbers, which are all each equal to an even number, as is below in consecutive order:

3 + 9 = 5 + 7 = 1 + 11 = **12**

5 + 9 = 3 + 11 = 7 + 7 = 1 + 13 = **14**

7 + 9 = 3 + 13 = 5 + 11 = **16**

3 + 15 = 7 + 11 = 5 + 13 = 1 + 17 **= 18**

11 + 9 = 3 + 17 = 7 + 13 = 1 + 19 **= 20**

13 + 9 = 3 + 19 = 5 + 17 = 11 + 11 **= 22**

3 + 21 = 11 + 13 = 5 + 19 = 7 + 17 = 1 + 23 **= 24**

17 + 9 = 3 + 23 = 7 + 19 = 13 + 13 **= 26**

19 + 9 = 5 + 23 = 11 + 17 **= 28**

5 + 25 = 13 + 17 = 11 + 19 = 7 + 23 = 1 + 29 **= 30**

23 + 9 = 3 + 29 = 13 + 19 = 1 + 31 **= 32**

7 + 27 = 17 + 17 = 3 + 31 = 5 + 29 = 11 + 23 = 17 + 17 **= 34**

3 + 33 = 17 + 19 = 5 + 31 = 7 + 29 = 13 + 23 **= 36**

29 + 9 = 7 + 31 = 19 + 19 = 1 + 37 **= 38**

31 + 9 = 3 + 37 = 11 + 29 = 17 + 23 **= 40**

3 + 39 = 19 + 23 = 5 + 37 = 11 + 31 = 13 + 29 = 1 + 41 **= 42**

5 + 39 = 3 + 41 = 7 + 37 = 13 + 31 = 1 + 43 **= 44**

37 + 9 = 3 + 43 = 5 + 41 = 17 + 29 = 23 + 23 **= 46**

3 + 45 = 5 + 43 = 7 + 41 = 11 + 37 = 17 + 31 = 19 + 29 = 1 + 47 **= 48**

41 + 9 = 3 + 47 = 7 + 43 = 13 + 37 = 19 + 31 **= 50**

43 + 9 = 5 + 47 = 11 + 41 = 23 + 29 **= 52**

5 + 49 = 7 + 47 = 11 + 43 = 13 + 41 = 17 + 37 = 23 + 31 = 1 + 53 **= 54**

47 + 9 = 3 + 53 = 13 + 43 = 19 + 37 **= 56**

3 + 55 = 29 + 29 = 5 + 53 = 11 + 47 = 17 + 41 = 29 + 29 **= 58**

5 + 55 = 29 + 31 = 7 + 53 = 13 + 47 = 17 + 43 = 19 + 41 = 23 + 37 = 1 + 59 **= 60**

53 + 9 = 3 + 59 = 19 + 43 = 31 + 31 = 1 + 61 = **62**

7 + 57 = 3 + 61 = 5 + 59 = 11 + 53 = 17 + 47 = 23 + 41 = **64**

11 + 55 = 5 + 61 = 7 + 59 = 13 + 53 = 19 + 47 = 23 + 43 = 29 + 37 = **66**

59 + 9 = 7 + 61 = 31 + 37 = 1 + 67 = **68**

61 + 9 = 3 + 67 = 11 + 59 = 17 + 53 = 23 + 47 = 29 + 41 = **70**

3 + 69 = 5 + 67 = 11 + 61 = 13 + 59 = 19 + 53 = 29 + 43 = 31 + 41 = 1 + 71 = **72**

5 + 69 = 37 + 37 = 3 + 71 = 7 + 67 = 13 + 61 = 31 + 43 = 37 + 37 = 1 + 73 = **74**

67 + 9 = 3 + 73 = 5 + 71 = 17 + 59 = 23 + 53 = 29 + 47 = **76**

3 + 75 = 37 + 41 = 5 + 73 = 7 + 71 = 11 + 67 = 31 + 47 = 37 + 41 = **78**

71 + 9 = 7 + 73 = 13 + 67 = 19 + 61 = 37 + 43 = 1 + 79 = **80**

73 + 9 = 3 + 79 = 11 + 71 = 23 + 59 = 29 + 53 = 41 + 41 = **82**

3 + 81 = 41 + 43 = 5 + 79 = 11 + 73 = 13 + 71 = 17 + 67 = 23 + 61 = 31 + 53 = 37 + 47 = 1+ 83 = **84**

5 + 81 = 43 + 43 = 3 + 83 = 7 + 79 = 13 + 73 = 19 + 67 = 43 + 43 = **86**

79 + 9 = 5 + 83 = 17 + 71 = 29 + 59 = 41 + 47 = **88**

3 + 87 = 7 + 83 = 11 + 79 = 17 + 73 = 19 + 71 = 23 + 67 = 29 + 61 = 31 + 59 = 37 + 53 = 43 + 47 = 1 + 89 = **90**

83 + 9 = 3 + 89 = 13 + 79 = 19 + 73 = 31 + 61 = 1 + 91 = **92**

7 + 87 = 47 + 47 = 5 + 89 = 11 + 83 = 23 + 71 = 41 + 53 = 47 + 47 = **94**

3 + 93 = 5 + 91 = 7 + 89 = 13 + 83 = 17 + 79 = 23 + 73 = 29 + 67 = 37 + 59 = 43 + 53 = **96**

89 + 9 = 7 + 91 = 19 + 79 = 31 + 67 = 37 + 61 = 1 + 97 = **98**

91 + 9 = 3 + 97 = 11 + 89 = 17 + 83 = 29 + 71 = 41 + 59 = 47 + 53 = **100**

3 + 99 = 5 + 97 = 11 + 91 = 13 + 89 = 19 + 83 = 23 + 79 = 29 + 73 = 31 + 71 = 41+ 61 = 43 + 59 = 1 + 101

= **102**

.

.

.

c) Every sum of 2 odd composites which are products of primes which are odd, is equal to the sum of 2 primes which are odd numbers, which are all each equal to an even number, as is below in consecutive order:

9 + 9 = 5 + 13 = 7 + 11 = 1 + 17 = **18**

9 + 15 = 5 + 19 = 7 + 17 = 11 + 13 = 1 + 23 = **24**

15 + 15 = 7 + 23 = 11 + 19 = 13 + 17 = 1 + 29 = **30**

9 + 25 = **7 + 27** = 17 + 17 = 3 + 31 = 5 + 29 = 11 + 23 = 17 + 17 = **34**

15 + 21 = 5 + 31 = 7 + 29 = 13 + 23 = 17 + 19 = **36**

15 + 25 = 3 + 37 = 11 + 29 = 17 + 23 = **40**

21 + 21 = 5 + 37 = 11 + 31 = 13 + 29 = 19 + 23 = 1 + 41 = **42**

9 + 35 = 3 + 41 = 7 + 37 = 13 + 31 = 1 + 43 = **44**

21 + 25 = 3 + 43 = 5 + 41 = 17 + 29 = 23 + 23 = **46**

9 + 39 = 5 + 43 = 7 + 41 = 11 + 37 = 17 + 31 = 19 + 29 = 1 + 47 = **48**

25 + 25 = 3 + 47 = 7 + 43 = 13 + 37 = 19 + 31 = **50**

25 + 27 = 5 + 47 = 11 + 41 =23 + 29 = **52**

27 + 27 = 7 + 47 = 11 + 43 = 13 + 41 = 17 + 37 = 23 + 31 = 1 + 53 = **54**

21 + 35 = 3 + 53 = 13 + 43 = 19 + 37 = **56**

9 + 49 = 29 + 29 = 5 + 53 = 11 + 47 = 17 + 41 = 29 + 29 = **58**

27 + 33 = 7 + 53 = 13 + 47 = 17 + 43 = 19 + 41 = 23 + 37 = 29 + 31 = 1 + 59 = **60**

27 + 35 = 31 + 31 = 3 + 59 = 19 + 43 = 1 + 61 = **62**

9 + 55 = 3 + 61 = 5 + 59 = 11 + 53 = 17 + 47 = 23 + 41 = **64**

33 + 33 = 5 + 61 = 7 + 59 = 13 + 53 = 19 + 47 = 23 + 43 = 29 + 37 = **66**

33 + 35 = 7 + 61 = 31 + 37 = 1 + 67 = **68**

35 + 35 = 3 + 67 = 11 + 59 = 17 + 53 = 23 + 47 = 29 + 41 = **70**

9 + 63 = 5 + 67 = 11 + 61 = 13 + 59 = 19 + 53 = 29 + 43 = 31 + 41 = 1 + 71 = **72**

35 + 39 = 3 + 71 = 7 + 67 = 13 + 61 = 31 + 43 = 37 + 37 = 1 + 73 = **74**

21 + 55 = 3 + 73 = 5 + 71 = 17 + 59 = 23 + 53 = 29 + 47 = **76**

39 + 39 = 5 + 73 = 7 + 71 = 11 + 67 = 31 + 47 = 37 + 41 = **78**

15 + 65 = 7 + 73 = 13 + 67 = 19 + 61 = 37 + 43 = 1 + 79 = **80**

25 + 57 = 41 + 41 = 3 + 79 = 11 + 71 = 23 + 59 = 29 + 53 = **82**

39 + 45 = 5 + 79 = 11 + 73 = 13 + 71 = 17 + 67 = 23 + 61 = 31 + 53 = 37 + 47 = 41 + 43 = 1 + 83 = **84**

9 + 77 = 43 + 43 = 3 + 83 = 7 + 79 = 13 + 73 = 19 + 67 = 43 + 43 = **86**

25 + 63 = 5 + 83 = 17 + 71 = 29 + 59 = 41 + 47 = **88**

45 + 45 = 7 + 83 = 11 + 79 = 17 + 73 = 19 + 71 = 23 + 67 = 29 + 61 = 31 + 59 = 37 + 53 = 43 + 47 = 1 + 89 = **90**

15 + 77 = 3 + 89 = 13 + 79 = 19 + 73 = 31 + 61 = 1 + 91 = **92**

45 + 49 = 5 + 89 = 11 + 83 = 23 + 71 = 41 + 53 = 47 + 47 = **94**

9 + 87 = 5 + 91 = 7 + 89 = 13 + 83 = 17 + 79 = 23 + 73 = 29 + 67 = 37 + 59 = 43 + 53 = **96**

49 + 49 = 7 + 91 = 19 + 79 = 31 + 67 = 37 + 61 = 1 + 97 = **98**

49 + 51 = 3 + 97 = 11 + 89 = 17 + 83 = 29 + 71 = 41 + 59 = 47 + 53 = **100**

51 + 51 = 5 + 97 = 11 + 91 = 13 + 89 = 19 + 83 = 23 + 79 = 29 + 73 = 31 + 71 = 41 + 61 = 43 + 59 = 1 +

101 = **102**

.

.

.

d) From (a), (b) & (c) above, we have the even numbers from 4 to 102 … composed as follows:

1) **4** = 2 + 2 = 1 + 3 (sum of 2 primes only)

2) **6** = 3 + 3 = 1 + 5 (sum of 2 primes only)

3) **8** = 3 + 5 = 1 + 7 (sum of 2 primes only)

4) **10** = 5 + 5 = 3 + 7 (sum of 2 primes only)

5) **12** = 5 + 7 = 1 + 11 = **3 + 9** (sum of 1 prime & 1 odd composite)

6) **14** = 3 + 11 = 7 + 7 = 1 + 13 = **5 + 9** (sum of 1 prime & 1 odd composite)

7) **16** = 3 + 13 = 5 + 11 = **7 + 9** (sum of 1 prime & 1 odd composite)

8) **18** = 5 + 13 = 7 + 11 = 1 + 17 = **3 + 15** (sum of 1 prime & 1 odd composite) = **9 + 9** (sum of 2 odd

   composites)

9) **20** = 3 + 17 = 7 + 13 = 1 + 19 = **11 + 9** (sum of 1 prime & 1 odd composite)

10) **22** = 3 + 19 = 5 + 17 = 11 + 11 = **13 + 9** (sum of 1 prime & 1 odd composite)

11) **24** = 5 + 19 = 7 + 17 = 11 + 13 = 1 + 23 = **3 + 21** (sum of 1 prime & 1 odd composite) = **9 + 15** (sum of

   2 odd composites)

12) **26** = 3 + 23 = 7 + 19 = 13 + 13 = **17 + 9** (sum of 1 prime & 1 odd composite)

13) **28** = 5 + 23 = 11 + 17 = **19 + 9** (sum of 1 prime & 1 odd composite)

14) **30** = 7 + 23 = 11 + 19 = 13 + 17 = 1 + 29 = **5 + 25** (sum of 1 prime & 1 odd composite) **= 15 + 15** (sum

of 2 odd composites)

15) **32** = 3 + 29 = 13 + 19 = 1 + 31 = **23 + 9** (sum of 1 prime & 1 odd composite)

16) **34** = 17 + 17 = 3 + 31 = 5 + 29 = 11 + 23 = 17 + 17 = **7 + 27** (sum of 1 prime & 1 odd composite) = **9 + 25** (sum of 2 odd composites)

17) **36** = 5 + 31 = 7 + 29 = 13 + 23 = 17 + 19 = **3 + 33** (sum of 1 prime & 1 odd composite) = **15 + 21** (sum of 2 odd composites)

18) **38** = 7 + 31 = 19 + 19 = 1 + 37 = **29 + 9** (sum of 1 prime & 1 odd composite)

19) **40** = 3 + 37 = 11 + 29 = 17 + 23 = **31 + 9** (sum of 1 prime & 1 odd composite) = **15 + 25** (sum of 2 odd composites)

20) **42** = 5 + 37 = 11 + 31 = 13 + 29 = 19 + 23 = 1 + 41 = **3 + 39** (sum of 1 prime & 1 odd composite) = **21 + 21** (sum of 2 odd composites)

21) **44** = 3 + 41 = 7 + 37 = 13 + 31 = 1 + 43 = **5 + 39** (sum of 1 prime & 1 odd composite) = **9 + 35** (sum of 2 odd composites)

22) **46** = 3 + 43 = 5 + 41 = 17 + 29 = 23 + 23 = **37 + 9** (sum of 1 prime & 1 odd composite) = **21 + 25** (sum of 2 odd composites)

23) **48** = 5 + 43 = 7 + 41 = 11 + 37 = 17 + 31 = 19 + 29 = 1 + 47 = **3 + 45** (sum of 1 prime & 1 odd composite) = **9 + 39** (sum of 2 odd composites)

24) **50** = 3 + 47 = 7 + 43 = 13 + 37 = 19 + 31 = **41 + 9** (sum of 1 prime & 1 odd composite) = **25 + 25** (sum of 2 odd composites)

25) **52** = 5 + 47 = 11 + 41 =23 + 29 = = **43 + 9** (sum of 1 prime & 1 odd composite) = **25 + 27** (sum of 2 odd composites)

26) **54** = 7 + 47 = 11 + 43 = 13 + 41 = 17 + 37 = 23 + 31 = 1 + 53 = **5 + 49** (sum of 1 prime & 1 odd

composite) = **27 + 27** (sum of 2 odd composites)

27) **56** = 3 + 53 = 13 + 43 = 19 + 37 = **47 + 9** (sum of 1 prime & 1 odd composite) = **21 + 35** (sum of 2 odd composites)

28) **58** = 29 + 29 = 5 + 53 = 11 + 47 = 17 + 41 = 29 + 29 = **3 + 55** (sum of 1 prime & 1 odd composite) = **9 + 49** (sum of 2 odd composites)

29) **60** = 7 + 53 = 13 + 47 = 17 + 43 = 19 + 41 = 23 + 37 = 29 + 31 = 1 + 59 = **5 + 55** (sum of 1 prime & 1 odd composite) = **27 + 33** (sum of 2 odd composites)

30) **62** = 3 + 59 = 19 + 43 = 31 + 31 = 1 + 61 = **53 + 9** (sum of 1 prime & 1 odd composite) = **27 + 35** (sum of 2 odd composites)

31) **64** = 3 + 61 = 5 + 59 = 11 + 53 = 17 + 47 = 23 + 41 = **7 + 57** (sum of 1 prime & 1 odd composite) = **9 + 55** (sum of 2 odd composites)

32) **66** = 5 + 61 = 7 + 59 = 13 + 53 = 19 + 47 = 23 + 43 = 29 + 37 = **11 + 55** (sum of 1 prime & 1 odd composite) = **33 + 33** (sum of 2 odd composites)

33) **68** = 7 + 61 = 31 + 37 = 1 + 67 = **59 + 9** (sum of 1 prime & 1 odd composite) = **33 + 35** (sum of 2 odd composites)

34) **70** = 3 + 67 = 11 + 59 = 17 + 53 = 23 + 47 = 29 + 41 = **61 + 9** (sum of 1 prime & 1 odd composite) = **35 + 35** (sum of 2 odd composites)

35) **72** = 5 + 67 = 11 + 61 = 13 + 59 = 19 + 53 = 29 + 43 = 31 + 41 = 1 + 71 = **3 + 69** (sum of 1 prime & 1 odd composite) = **9 + 63** (sum of 2 odd composites)

36) **74** = 3 + 71 = 7 + 67 = 13 + 61 = 31 + 43 = 37 + 37 = 1 + 73 = **5 + 69** (sum of 1 prime & 1 odd composite) = **35 + 39** (sum of 2 odd composites)

37) **76** = 3 + 73 = 5 + 71 = 17 + 59 = 23 + 53 = 29 + 47 = **67 + 9** (sum of 1 prime & 1 odd composite) = **21**

+ **55** (sum of 2 odd composites)

38) **78** = 5 + 73 = 7 + 71 = 11 + 67 = 31 + 47 = 37 + 41 = **3 + 75** (sum of 1 prime & 1 odd composite) = **39**

+ **39** (sum of 2 odd composites)

39) **80** = 7 + 73 = 13 + 67 = 19 + 61 = 37 + 43 = 1 + 79 = **71 + 9** (sum of 1 prime & 1 odd composite) = **15**

+ **65** (sum of 2 odd composites)

40) **82** = 3 + 79 = 11 + 71 = 23 + 59 = 29 + 53 = 41 + 41 = **73 + 9** (sum of 1 prime & 1 odd composite) = **25**

+ **57** (sum of 2 odd composites)

41) **84** = 5 + 79 = 11 + 73 = 13 + 71 = 17 + 67 = 23 + 61 = 31 + 53 = 37 + 47 = 41 + 43 = 1 + 83 = **3 + 81**

(sum of 1 prime & 1 odd composite) = **39 + 45** (sum of 2 odd composites)

42) **86** = 43 + 43 = 3 + 83 = 7 + 79 = 13 + 73 = 19 + 67 = 43 + 43 = **5 + 81** (sum of 1 prime & 1 odd

composite) = **9 + 77** (sum of 2 odd composites)

43) **88** = 5 + 83 = 17 + 71 = 29 + 59 = 41 + 47 = **79 + 9** (sum of 1 prime & 1 odd composite) = **25 + 63** (sum

of 2 odd composites)

44) **90** = 7 + 83 = 11 + 79 = 17 + 73 = 19 + 71 = 23 + 67 = 29 + 61 = 31 + 59 = 37 + 53 = 43 + 47 = 1 + 89

= **3 + 87** (sum of 1 prime & 1 odd composite) = **45 + 45** (sum of 2 odd composites)

45) **92** = 3 + 89 = 13 + 79 = 19 + 73 = 31 + 61 = 1 + 91 = **83 + 9** (sum of 1 prime & 1 odd composite) = **15**

+ **77** (sum of 2 odd composites)

46) **94** = 5 + 89 = 11 + 83 = 23 + 71 = 41 + 53 = 47 + 47 = **7 + 87** (sum of 1 prime & 1 odd composite) = **45**

+ **49** (sum of 2 odd composites)

47) **96** = 5 + 91 = 7 + 89 = 13 + 83 = 17 + 79 = 23 + 73 = 29 + 67 = 37 + 59 = 43 + 53 = **3 + 93** (sum of 1

prime & 1 odd composite) = **9 + 87** (sum of 2 odd composites)

48) **98** = 7 + 91 = 19 + 79 = 31 + 67 = 37 + 61 = 1 + 97 = **89 + 9** (sum of 1 prime & 1 odd composite) = **49**

+ 49 (sum of 2 odd composites)

49) **100** = 3 + 97 = 11 + 89 = 17 + 83 = 29 + 71 = 41 + 59 = 47 + 53 = **91 + 9** (sum of 1 prime & 1 odd

composite) = **49 + 51** (sum of 2 odd composites)

50) **102** = 5 + 97 = 11 + 91 = 13 + 89 = 19 + 83 = 23 + 79 = 29 + 73 = 31 + 71 = 41 + 61 = 43 + 59 = 1 +

101 = **3 + 99** (sum of 1 prime & 1 odd composite) = **51 + 51** (sum of 2 odd composites)

.

.

.

(The above is only a partial or incomplete listing of sums of 1 prime & 1 odd composite, and, sums of 2 odd composites, each of which is equal to the sum of 2 primes as well as an even number. For example, in the list of compositions for the even numbers 4 to 102 … above, in Item (48), we could also have other "combinations" such as: **98** = 7 + 91 = 19 + 79 = 31 + 67 = 37 + 61 = 1 + 97 = **25 + 73** (sum of 1 prime & 1 odd composite) = **21 + 77** (sum of 2 odd composites), etc., in Item (49), we could also have other "combinations" such as: **100** = 3 + 97 = 11 + 89 = 17 + 83 = 29 + 71 = 41 + 59 = 47 + 53 = **31 + 69** (sum of 1 prime & 1 odd composite) = **45 + 55** (sum of 2 odd composites), etc., and, in Item (50), we could also have other "combinations" such as: **102** = 5 + 97 = 11 + 91 = 13 + 89 = 19 + 83 = 23 + 79 = 29 + 73 = 31 + 71 = 41 + 61 = 43 + 59 = 1 + 101 = **17 + 85** (sum of 1 prime & 1 odd composite) = **21 + 81** (sum of 2 odd composites), etc. That is, there are more "combinations" than those shown in the above listing.)

In (d) above, in the list of compositions for the 50 consecutive even numbers 4 to 102 …, the even numbers 4, 6, 8 and 10 are only formed through the summing of 2 primes and not at all through the summing of 1 prime and 1 odd composite, or, the summing of 2 odd composites, which are impossibilities here. These sums of 2 primes are present (always present) throughout the whole list of compositions, from 4 right through to 102, while this is not the case for the sums of 1 prime and 1 odd composite, and, the sums of 2 odd composites.

We reason here by the process of elimination, through analysing the information in (d) above which pertains to the compositions of the 50 consecutive even numbers 4 to 102 … taken from the infinite list of even numbers. We stated at the beginning the following about the even numbers after 2:-

Firstly, every even number after 2 is:

A) The sum of 2 odd numbers.

   (Every odd number is either a prime which is odd or a composite - product of primes which are odd.

   Notably, every prime with the exception of 2 is an odd number.)

Secondly, every even number after 2 is also (the below-mentioned is the logical consequence of (A) above):

1) The sum of 2 primes which are odd.

2) And/or the sum of 1 prime which is odd and 1 odd composite whose prime factors are odd.

3) And/or the sum of 2 odd composites whose prime factors are odd.

Evidently, at least 1 of (1), (2) & (3) above has to be the "atom" or building-block of the even numbers. In (d) above, we observe the following:-

i)   All the 50 consecutive even numbers 4 to 102 … in (d) above taken from the infinite list of even numbers are sums of 2 primes.

ii)   It is impossible for each of the even numbers 4, 6, 8 & 10 in (d) above to be the sum of 1 prime which is odd and 1 odd composite whose prime factors are odd.

iii)   It is impossible for each of the even numbers 4, 6, 8, 10, 12, 14, 16, 20, 22, 26, 28, 32 & 38 in (d) above to be the sum

        of 2 odd composites whose prime factors are odd.

It is evident from (i), (ii) & (iii) above that neither (2) nor (3) can be the "atom" or building-block of the even numbers since they are "incomplete". As (1) - the sum of 2 primes which are odd - is "complete", i.e., always present in the 50 consecutive even numbers 4 to 102 … in (d) above, unlike (2) & (3), it evidently is the "atom" or building-block of the even numbers. That is, every even number after 2 is evidently the sum of 2 primes which are odd. In fact, a distributed computer search completed in 2008 at the University of Aveiro, Portugal, had verified this for all even numbers up to 12×10^{17}, which is not a small list. Definitely, due respectively to (ii) & (iii) above, we cannot say that every even number after 2 is the sum of 1 prime which is odd and 1 odd composite whose prime factors are odd, or, every even number after 2 is the sum of 2 odd composites whose prime factors are odd.

By the above lemma and corollary, the infinitudes of the primes, even numbers and odd numbers indeed imply that there are an infinite number of sums of 2 primes which are odd numbers, which are each equal to an even number. As the sums of 2 primes which are odd numbers are evidently the "atoms" or building-blocks of the even numbers, it also implies that they are infinite, since the even numbers are infinite.

Hypothetically, if on the other hand just 1 of the 3 items stated above, primes, even numbers and odd numbers, were finite, the above-said sums of 2 primes which are odd numbers, each of which is equal to an even number, would be finite. The primes, even numbers and odd numbers are evidently intricately linked, with the primes playing the part of building-blocks of both the even and odd numbers through various "combinations" as is described below. However, as the primes, even numbers and odd numbers are intricately linked, the finiteness (or, infinity) of any 1 of them implies the finiteness (or, infinity) of the other 2, and vice versa. These 3 items are evidently "close comrades-in-arm" working together to give special meaning to the integers. As these 3 are all infinite, it indeed implies that there is an infinitude of even numbers which are infinitely the sums of 2 primes that are odd and infinite.

Argument 2:-

Lemma:

According to the precepts of fractal geometry and group theory, symmetry is a very important, intrinsic part of nature. There is symmetry all around us and within us. There is evident symmetry in human bodies, the structures of viruses and bacteria, polymers and ceramic materials, the permutations of numbers, the universe and many others, even the movements of prices in financial markets, the growths of populations, the sound of music, the flow of blood through our circulatory system, the behaviour of people en masse, etc. In other words, regularity, pattern, order, uniformity or symmetry is evident everywhere.

The above-mentioned most basic, always present sums of 2 primes, each of which is equal to an even number, which are evidently the "atoms" or building-blocks of the even numbers, are characterised by the feature of symmetry (in 2008, a distributed computer search ran by Tomas Oliveira e Silva, a researcher at the University of Aveiro, Portugal, had verified that all even numbers up to 12×10^{17}, which is no small list of numbers, are sums of 2 primes, a regularity, uniformity, order, pattern, symmetry). Thus, by the above lemma, every even number after 2 is naturally or inherently the sum of 2 primes, i.e., there is an infinitude of sums of 2 primes which are each equal to an even number.

Hence, the confirmation of the following generalisation pertaining to the integers, whereby it is indeed evident that the primes play a very important role:

Let a prime = p, &, a composite = c = p x p ..... .

a) Every even number after $2 = p + p = {}^*c = {}^*p \times p$ .... &/V $= c + p = (p \times p$ ....$) + p$

&/V $= c + c = (p \times p$ ....$) + (p \times p$ ....$)$ (in $^*c = {}^*p \times p$ .... here, which is an even

composite, 1 or more of the p's are 2, the only even prime, e.g., $6 = 2 \times 3$, $8 = 2 \times$

2×2, $10 = 2 \times 5$, $18 = 2 \times 3 \times 3$, $20 = 2 \times 2 \times 5$, $24 = 2 \times 2 \times 2 \times 3$, etc.)

b) Every odd number = p V c = p x p .... (in c = p x p .... here, which is an odd

composite, like the c = p x p ....'s in (a) above, all the p's are odd, e.g., $9 = 3 \times 3$, 15

$= 3 \times 5$, $21 = 3 \times 7$, $25 = 5 \times 5$, $63 = 3 \times 3 \times 7$, $99 = 3 \times 3 \times 11$, etc.)

It is easy to see that the Goldbach conjecture is valid, i.e., every even number after 2 is the sum of 2 primes.

PART 4

Theorem:- Every even number after 2 is the sum of 2 primes.

Solution:-
The prime numbers are evidently the atoms or building-blocks of the integers. The integers are either primes (not divisible by other integers except 1) or composites (divisible by other integers, e.g., the prime numbers), and, even (the sums of 2 primes as conjectured by Goldbach) or odd (primes, or, composites whereby they are divisible by prime factors). Therefore, to determine whether the conjecture that every even number (except the number 2) is the sum of 2 primes is true, it would be appropriate to analyse the evident atoms or building-blocks of the even numbers, viz., the prime numbers. For the solution to this conjecture we note that the primes (vide Euclid's proof) and the even numbers are infinite, which implies that this conjecture should be true.

We here analyse the "behaviour" of the first 2,400 consecutive prime numbers (divided into 12 batches of consecutive primes, each subsequent batch with an increment of 200 primes), leaving out 2 (because it is an even prime) and commencing with 3, which is the 2nd. consecutive prime, the latter to be the first prime in our list of 2,400 consecutive primes (3 to 21,391), as follows:-

(1) 200 Consecutive Primes From 3 To 1,229
    (a) Even numbers (obtained by summing of 2 primes) = 6 to 2,458
    (b) No. of even numbers = 1,227
    (c) No. of primes = 200
    (d) Average no. of even numbers "generated" by each of these 200 consecutive primes =
        1,227 ÷ 200 = **6.14**
    (e) No. of summings of 2 primes/permutations (3 + 3, 3 + 5, 3 + 7, 3 + 11, ...
        etc.) for these 200 primes = 200 x 200 = 40,000
    (f) Average no. of summings of 2 primes/permutations for each of the 1,227 even
        numbers = 40,000 ÷ 1,227 = **32.60**

(2) 400 Consecutive Primes From 3 To 2,749
    (a) Even numbers (obtained by summing of 2 primes) = 6 to 5,498
    (b) No. of even numbers = 2,747
    (c) No. of primes = 400
    (d) Average no. of even numbers "generated" by each of these 400 consecutive primes =
        2,747 ÷ 400 = **6.87**
    (e) No. of summings of 2 primes/permutations (3 + 3, 3 + 5, 3 + 7, 3 + 11, ...
        etc.) for these 400 primes = 400 x 400 = 160,000
    (f) Average no. of summings of 2 primes/permutations for each of the 2,747 even
        numbers = 160,000 ÷ 2,747 = **58.25**

(3) 600 Consecutive Primes From 3 To 4,421
    (a) Even numbers (obtained by summing of 2 primes) = 6 to 8,842
    (b) No. of even numbers = 4,419
    (c) No. of primes = 600
    (d) Average no. of even numbers "generated" by each of these 600 consecutive primes =
        4,419 ÷ 600 = **7.37**
    (e) No. of summings of 2 primes/permutations (3 + 3, 3 + 5, 3 + 7, 3 + 11, ...
        etc.) for these 600 primes = 600 x 600 = 360,000
    (f) Average no. of summings of 2 primes/permutations for each of the 4,419 even

numbers = 360,000 ÷ 4,419 = **81.47**

(4) <u>800 Consecutive Primes From 3 To 6,143</u>
   (a) Even numbers (obtained by summing of 2 primes) = 6 to 12,286
   (b) No. of even numbers = 6,141
   (c) No. of primes = 800
   (d) Average no. of even numbers "generated" by each of these 800 consecutive primes = 6,141 ÷ 800 = **7.68**
   (e) No. of summings of 2 primes/permutations (3 + 3, 3 + 5, 3 + 7, 3 + 11, ... etc.) for these 800 primes = 800 x 800 = 640,000
   (f) Average no. of summings of 2 primes/permutations for each of the 6,141 even numbers = 640,000 ÷ 6,141 = **104.22**

(5) <u>1,000 Consecutive Primes From 3 To 7,927</u>
   (a) Even numbers (obtained by summing of 2 primes) = 6 to 15,854
   (b) No. of even numbers = 7,925
   (c) No. of primes = 1,000
   (d) Average no. of even numbers "generated" by each of these 1,000 consecutive primes = 7,925 ÷ 1,000 = **7.93**
   (e) No. of summings of 2 primes/permutations (3 + 3, 3 + 5, 3 + 7, 3 + 11, ... etc.) for these 1,000 primes = 1,000 x 1,000 = 1,000,000
   (f) Average no. of summings of 2 primes/permutations for each of the 7,925 even numbers = 1,000,000 ÷ 7,925 = **126.18**

(6) <u>1,200 Consecutive Primes From 3 To 9,739</u>
   (a) Even numbers (obtained by summing of 2 primes) = 6 to 19,478
   (b) No. of even numbers = 9,737
   (c) No. of primes = 1,200
   (d) Average no. of even numbers "generated" by each of these 1,200 consecutive primes = 9,737 ÷ 1,200 = **8.11**
   (e) No. of summings of 2 primes/permutations (3 + 3, 3 + 5, 3 + 7, 3 + 11, ... etc.) for these 1,200 primes = 1,200 x 1,200 = 1,440,000
   (f) Average no. of summings of 2 primes/permutations for each of the 9,737 even numbers = 1,440,000 ÷ 9,737 = **147.89**

(7) <u>1,400 Consecutive Primes From 3 To 11,677</u>
   (a) Even numbers (obtained by summing of 2 primes) = 6 to 23,354
   (b) No. of even numbers = 11,675
   (c) No. of primes = 1,400
   (d) Average no. of even numbers "generated" by each of these 1,400 consecutive primes = 11,675 ÷ 1,400 = **8.34**
   (e) No. of summings of 2 primes/permutations (3 + 3, 3 + 5, 3 + 7, 3 + 11, ... etc.) for these 1,400 primes = 1,400 x 1,400 = 1,960,000
   (f) Average no. of summings of 2 primes/permutations for each of the 11,675 even numbers = 1,960,000 ÷ 11,675 = **167.88**

(8) <u>1,600 Consecutive Primes From 3 To 13,513</u>
   (a) Even numbers (obtained by summing of 2 primes) = 6 to 27,026
   (b) No. of even numbers = 13,511

(c) No. of primes = 1,600

(d) Average no. of even numbers "generated" by each of these 1,600 consecutive primes
= 13,511 ÷ 1,600 = **8.44**

(e) No. of summings of 2 primes/permutations (3 + 3, 3 + 5, 3 + 7, 3 + 11, ...
etc.) for these 1,600 primes = 1,600 x 1,600 = 2,560,000

(f) Average no. of summings of 2 primes/permutations for each of the 13,511
even numbers = 2,560,000 ÷ 13,511 = **189.48**

(9) 1,800 Consecutive Primes From 3 To 15,413

(a) Even numbers (obtained by summing of 2 primes) = 6 to 30,826

(b) No. of even numbers = 15,411

(c) No. of primes = 1,800

(d) Average no. of even numbers "generated" by each of these 1,800 consecutive primes
= 15,411 ÷ 1,800 = **8.56**

(e) No. of summings of 2 primes/permutations (3 + 3, 3 + 5, 3 + 7, 3 + 11, ...
etc.) for these 1,800 primes = 1,800 x 1,800 = 3,240,000

(f) Average no. of summings of 2 primes/permutations for each of the 15,411
even numbers = 3,240,000 ÷ 15,411 = **210.24**

(10) 2,000 Consecutive Primes From 3 To 17,393

(a) Even numbers (obtained by summing of 2 primes) = 6 to 34,786

(b) No. of even numbers = 17,391

(c) No. of primes = 2,000

(d) Average no. of even numbers "generated" by each of these 2,000 consecutive
primes = 17,391 ÷ 2,000 = **8.70**

(e) No. of summings of 2 primes/permutations (3 + 3, 3 + 5, 3 + 7, 3 + 11, ...
etc.) for these 2,000 primes = 2,000 x 2,000 = 4,000,000

(f) Average no. of summings of 2 primes/permutations for each of the 17,391
even numbers = 4,000,000 ÷ 17,391 = **230.00**

(11) 2,200 Consecutive Primes From 3 To 19,427

(a) Even numbers (obtained by summing of 2 primes) = 6 to 38,854

(b) No. of even numbers = 19,425

(c) No. of primes = 2,200

(d) Average no. of even numbers "generated" by each of these 2,200 consecutive
primes = 19,425 ÷ 2,200 = **8.83**

(e) No. of summings of 2 primes/permutations (3 + 3, 3 + 5, 3 + 7, 3 + 11, ...
etc.) for these 2,200 primes = 2,200 x 2,200 = 4,840,000

(f) Average no. of summings of 2 primes/permutations for each of the 19,425
even numbers = 4,840,000 ÷ 19,425 = **249.16**

(12) 2,400 Consecutive Primes From 3 To 21,391

(a) Even numbers (obtained by summing of 2 primes) = 6 to 42,782

(b) No. of even numbers = 21,389

(c) No. of primes = 2,400

(d) Average no. of even numbers "generated" by each of these 2,400 consecutive
primes = 21,389 ÷ 2,400 = **8.91**

(e) No. of summings of 2 primes/permutations (3 + 3, 3 + 5, 3 + 7, 3 + 11, ...

etc.) for these 2,400 primes = 2,400 x 2,400 = 5,760,000
  (f) Average no. of summings of 2 primes/permutations for each of the 21,389
      even numbers = 5,760,000 ÷ 21,389 = **269.30**

There would evidently be more and more profuse repetitions and overlaps of the even numbers "generated" by the primes the higher up the infinite list of prime numbers we go, which is significant.

We compare all the (d)s and (f)s in (1) to (12) above, which is as follows:-

  (d) Average no. of even numbers "generated" by each of the consecutive primes in (1) to
      (12) above, as follows according to the listings (1) to (12):

      (1) **6.14**, (2) **6.87**, (3) **7.37**, (4) **7.68**, (5) **7.93**, (6) **8.11**, (7) **8.34**, (8) **8.44**, (9) **8.56**, (10) **8.70**,
      (11) **8.83**, (12) **8.91**

  (f) Average no. of summings of 2 primes/permutations for each of the even numbers
      in (1) to (12) above, as follows according to the listings (1) to (12):

      (1) **32.60**, (2) **58.25**, (3) **81.47**, (4) **104.22**, (5) **126.18**, (6) **147.89**, (7) **167.88**, (8) **189.48**, (9) **210.24**,
      (10) **230.00**, (11) **249.16**, (12) **269.30**

The following is evident from the above information:-

  (A): (d) Average no. of even numbers "generated" by each of the consecutive primes in
           the above 12 listings increases continually all the way from the list: (1) 200
           Consecutive Primes From 3 To 1,229 to the list: (12) 2,400 Consecutive Primes
           From 3 To 21,391, from **6.14** even numbers per prime number in List (1) to **8.91**
           even numbers per prime number in List (12).
  (B): (f) Average no. of summings of 2 primes/permutations for each of the even numbers
           in the above 12 listings increases continually all the way from the list: (1) 200
           Consecutive Primes From 3 To 1,229 to the list: (12) 2,400 Consecutive Primes
           From 3 To 21,391, from **32.60** number of summings of 2 primes/permutations per
           even number in List (1) to **269.30** number of summings of 2 primes/permutations
           per even number in List (12).

Argument 1:
Lemma: According to the principle of complete induction in set theory, if a set of natural numbers contains 1 and, for each n, it contains $n +$
1 whenever it contains all numbers less than $n + 1$, then it must contain every natural number, e.g., complete induction proves that every
natural number is a product of primes.

By induction, we now deduce the following:

The larger the list of consecutive primes becomes, the greater would be the average number of even numbers "generated" by each of the
primes in the list of consecutive primes (inferred from (A) above).

The larger the list of consecutive primes becomes, the greater would be the average number of summings of 2 primes/permutations for each of the even numbers in the infinite list of even numbers (inferred from (B) above).

Furthermore, the Goldbach conjecture had been tested and found to be correct for every even number up to 12×10^{17}, which is not a small list, by a distributed computer search carried out at the University of Aveiro, Portugal, in 2008.

As the primes and the even numbers are infinite, by the above lemma and all the above deductions and information, it could be inferred that the increases stated in (A) and (B) above, with the even numbers each being the sum of 2 primes, continue to infinity, i.e., the Goldbach conjecture becomes stronger and stronger the higher up the infinite list of prime numbers/even numbers we go - all the way to infinity.

The validity of the Goldbach conjecture is thereby confirmed - every even number after 2 is the sum of 2 primes.

Argument 2:
Next, we resort to the argument by contradiction. The above deduction would be reversed if, e.g., the following takes place (which is the reversal of the above-mentioned information):

(A): (d) Average no. of even numbers "generated" by each of the consecutive primes in the above 12 listings decreases continually all the way from the list: (1) 200 Consecutive Primes From 3 To 1,229 to the list: (12) 2,400 Consecutive Primes From 3 To 21,391, from **8.91** even numbers per prime number in List (1) to **6.14** even numbers per prime number in List (12).

(B): (f) Average no. of summings of 2 primes/permutations for each of the even numbers in the above 12 listings decreases continually all the way from the list: (1) 200 Consecutive Primes From 3 To 1,229 to the list: (12) 2,400 Consecutive Primes From 3 To 21,391, from **269.30** number of summings of 2 primes/permutations per even number in List (1) to **32.60** number of summings of 2 primes/permutations per even number in List (12).

If this reversed state happens, the implication is that there would reach a point when there are no more batches of 2 prime numbers summing together to form even numbers, in which case the Goldbach conjecture would be false. Evidently this would happen when the prime numbers are finite. As the prime numbers are infinite (as Euclid had proved long ago) this would never happen.

Since the above information indicate otherwise, and, the prime numbers are infinite, we accept the above induction/deduction and infer that the Goldbach conjecture could not be false, i.e., the Goldbach conjecture is true, and, every even number (except 2) is indeed the sum of 2 prime numbers. This concludes the argument by contradiction.

Thus, by both induction and contradiction or *reductio ad absurdum* the validity of the Goldbach conjecture is proved.

CONCLUSION

It is evident here that the Goldbach conjecture could be approached in a number of different ways; a number of methods have been adopted in this chapter in proving the Goldbach conjecture.

The inductive method, which is a well-established proof, is one of the methods utilised. The following lends support to this inductive argument for the Goldbach conjecture: (a) The characteristic of a mountain or infinite volume of sand is reflected in the characteristic of some grains of sand found there so that studying the characteristic of some grains of sand found there is enough for deducing the characteristic of the mountain or infinite volume of sand, to ascertain the quality of a batch of products it is only necessary to inspect some carefully selected samples from that batch of products and not everyone of the products and to carry out a population census, i.e., find out the characteristics of a population, it is only necessary to carry out a survey on some carefully selected respondents and not the whole population; in like manner, by the same principle, we just need to study a carefully selected list of even numbers, find out whether they are all sums of 2 primes and deduce by induction whether all even numbers after this list would also be sums of 2 primes - this act is rather like extrapolation. (For example, a distributed computer search completed in 2008 at the University of Aveiro, Portugal, had confirmed that every even number up to 12×10^{17}, which is no small list of numbers, is the sum of 2 primes. By the principle of induction in this case we could deduce that all the even numbers after 12×10^{17} would also be sums of 2 primes.) (b) Thus, in this way every even number after 2 could be reasonably proved to be the sum of 2 primes. In fact, induction plays an important part in a number of the arguments.

The other argument used to prove the conjecture is the indirect (reductio ad absurdum) method, which had been used by Euclid and other mathematicians after him. Logically, 1 or 2 examples of "contradiction" should be sufficient proof of infinity, for it does not make sense to have a need for an infinite number of cases of "contradiction", as our proof would then have to be infinitely and impossibly long, an absurdity. This method of proof is "proof by implication" as a result of "contradiction" - which is a "short-cut" and smart way in proving infinity, instead of "proving infinity by counting to infinity", which is ludicrous, and, impossible. Hence, 1 or 2 cases of "contradiction" should be sufficient for implying that there would be an infinitude of even numbers which are sums of 2 primes, which of course also tacitly implies that there would be an infinitude of the number of cases of such "contradiction". (Euclid evidently had this logical point in mind when he formulated the indirect (reductio ad absurdum) proof of the infinity of the primes.) This method of proof had been cleverly used by a number of mathematicians, not the least by the great German mathematician, David Hilbert. For example, Hilbert had used an indirect method (the "reductio ad absurdum" proof) to prove Gordan'1s Theorem without having to show an actual "construction", a proof which had been accepted by his peers.

There is also the involvement of concepts from set theory, group theory, geometry, etc.

One important query here, which many might not have considered: What if the list of prime numbers is not infinite? Of course, if that is the case, the Goldbach conjecture would be false. It would then have been absurd for the Goldbach conjecture to have been conceived at all. However, the list of primes is infinite (vide Euclid's proof). This gives credence to the Goldbach conjecture.

A very important related point, in fact a most important point, must be highlighted here. If the Goldbach conjecture were indeed false, there must be an ultimate (largest) even number which is (and must necessarily be) the result of the summation of 2 primes that are each the largest existing prime. It must be noted that this is actually an impossibility, as there can never be a largest existing prime - by Euclid's proof, the primes are infinite (refer to Part 2, Argument 3 above). Hence, the Goldbach conjecture cannot be false, and, by both *reduction ad absurdum* (contradiction), and, induction (wherein all even numbers up to 12×10^{17}, not a small list, had been confirmed to be sums of 2 primes), has to be true.

Another important point is that the Goldbach conjecture becomes evidently stronger and stronger the higher up the infinite list of prime numbers/even numbers we go, as has been shown above. Thus, by implication, induction, extrapolation, it could be concluded that the Goldbach conjecture is valid - that every even number after 2 is the sum of 2 primes.

So far, there has been no indication or confirmation at all that the number of even numbers after the number 2 which are each the sum of 2 primes is finite and the largest existing even number which is the sum of 2 primes has not been found and confirmed. (This would of course be the case if the Goldbach conjecture is true.) On the other hand, practically everyone could intuit that the list of even numbers after the number 2 which are each the sum of 2 primes is infinite. Besides, the evidence, as shown in this chapter, is overwhelmingly in support of the infinity of this list.

We have no other more logical choice but to take the stand that every even number after the number 2 is the sum of 2 prime numbers.

In conclusion, we state that the Goldbach conjecture is true - every even number after the number 2 is indeed the sum of 2 primes.

8 EPILOGUE

Though a researcher may be confident that he has the solution to any of these outstanding prime number problems, he has to convince the mathematical community that it is the correct solution. He normally submits the solution to a mathematics journal for review by an expert or experts and if the solution is approved and accepted by these experts it would be published in the mathematics journal wherein other mathematicians would have access to the solution.

However, convincing the mathematical community is not likely to be an easy task. Many mathematics journals apparently refuse to consider papers on any of these famous prime number problems for in the past too many wrong papers, many of which were deemed to have been written by cranks, had been submitted. Not to despair, for the researcher could still post his solution online and he could get many approvals and support if the solution is correct.

Bibliography

[1] D. Burton, 1980, Elementary Number Theory, Allyn & Bacon

[2] R. Courant and H. Robbins, revised by I. Stewart, 1996, What Is Mathematics? An Elementary Approach to Ideas and Methods, Oxford University Press

[3] M. J. Feigenbaum, 1980, Universal Behavior In Nonlinear Systems, Los Alamos Science, No. 1

[4] G. H. Hardy and E. M. Wright, 1979, An Introduction To Theory Of Numbers, Oxford, England: Clarendon Press

[5] D. H. Lehmer, 1914, List Of Prime Numbers From 1 To 10,006,721, Publication No. 165, Carnegie Institution of Washington, Washington, D.C.

[6] M. E. Lines, 1986, A Number For Your Thoughts, Adam Hilger

[7] B. B. Mandelbrot, 1977, The Fractal Geometry Of Nature, W. H. Freeman

[8] R. McWeeny, 2002, Symmetry: An Introduction To Group Theory And Its Applications, Dover

[9] J. M. T. Thompson and H. B. Stewart, 1986, Nonlinear Dynamics And Chaos, New York: John Wiley

94

www.ingramcontent.com/pod-product-compliance
Lightning Source LLC
Chambersburg PA
CBHW081012170526
45158CB00010B/3010